DEDICATED TO

My wife Ruth, and our family
and, to my old friend, Roland W. Tapp, Th,D

Support and encouragement was truly appreciated.

By the same author

"Those Were The Days"

“COME, *FLY WITH ME*”
Mark A. “Doc” Savage

Front cover design: Robert Tourt

Photography: Russell Savage

Library of Congress Cataloging-in-Publication Data
LC number 90-91732

ISBN 1-878455-02-8

Published by: **MARKAS PUBLISHING**
P.O. Box 415
Dublin, OH 43017

Contents

Preface

It has often been said, "Flying is basically hours and hours of boredom, punctuated by seconds of shear terror." For many, these words speak for themselves. Despite the seemingly action packed pages herein, there have been thousands of hours, and years on end, where absolutely nothing has ever happened. Everything has been strictly routine.

Throughout this book, you may note the reference to gliders and sailplanes. Gliders were used during WW II for invasion purposes. We are now operating and flying sailplanes rather than gliders. The higher performance makes the difference. Even so, the name glider has stuck in the minds of everyone even to this time.

After 18,260 hours of all types of flying, I have on occasions experienced some of those tense moments. But, they really have been few and far between.

It is only natural for a person to recall the more vivid details of one's life.

The following stories are just a few of the many chapters of my life. I hope you will enjoy sharing these real true adventures with me.

Chapter I
The Crop Duster

I was in my second year at KSU and the spring quarter had just started. My wife Ruth, knew how much I had really missed flying, yet she was quite tolerant about my constant bitching and grumbling. I liked school now, especially since I was enrolled in the Kent State University's brand new aviation technology department. This program kept me more than just interested. The courses were OK, but . . . I sure did miss flying.

Although I was not flying at this time, I tried to go out to the airport just to be near airplanes. Late one afternoon, I was approached by a guy named Dick Baker who appeared to me to be a rather young kid. Well, at least he was a bit younger than me.

Some how, or some way, Dick found out about me, and my former military flying. "Say, Mark, I understand you flew in the military?"

"That's right," I replied wondering just what all this conversation was leading up to.

"Well, Mark, I have a ship, and I need a pilot to fly it for me."

I just about leaped out of my boots at this offer, but I held back cautiously and said, "That's nice, but just what kind of a ship is it?"

A WW II Stearman similar to this one before being converted over to become a crop duster/sprayer. Photo and Stearman model: Courtesy of Charles R. Moor.

"Why, it's an ex-military plane, a surplus WW II trainer, called a Stearman PT-17," he explained.

I had never flown that particular ship while I was on active duty in the military. My very first military ship was known as a Fairchild PT-19.

"OK, I'm interested," I said. "So, what do you want me to do?"

Dick replied, "How about taking a spin around the field and shoot a few landings? Get the feel of my ship, then we will simulate crop-dusting before we actually go out on the job site."

I didn't think much about being a crop duster, but the opportunity to fly again was more than I could resist. "OK, I'll take it out and be back in about a half hour or so," I replied.

"Great," said Dick, "I'll be waiting for you right here at the airport."

I had heard a lot about this fierce ground-looping trainer known as: "the Yellow Peril." I planned to be extra cautious and I wasn't about to let it give me any problems.

After I successfully completed the three take-offs and landings, I taxied back to the flight line exuberant about being back in the air again.

"You did great, Mark. Now, do you want the job?" asked Dick. Job, what job? I was almost ready to pay him for getting me back up in the air again.

"When do you want me to start, and by the way, how much pay is involved with this job?" I asked.

He could easily see how excited I was about flying, and pretended not to hear my questions.

"I want you to come back tomorrow so you can fly again, only this time I want you to simulate crop-dusting for practice."

"Oh, by the way. As far as pay is concerned, we won't get paid until the job is done. The pay is based on the actual acres we do."

Finally, Dick said, "Don't worry, you should make between $300-500 dollars."

WOW! $500! That's a lot of money. I've been living on a

mere $70 a month. Going to school on the GI Bill had been real tough.

I got back home to the mobile trailer home and told Ruth all about the good news and particularly about my getting back to flying once again. Believe me, she wasn't too happy about my new found field of work.

"For one thing dear," she said, "what would happen to me and the baby if you get hurt, or even killed?"

"Aw, that's nuts, Ruth. Nothing will happen."

After my simple speech, it seemed to me as though her fears were somewhat allayed.

I went to bed that night quite excited over the prospect of becoming a crop duster and making all that big money to boot.

That next afternoon at the airport I greeted Dick, and he proposed we go to a larger field just north of here, and make a few practice runs. Everything went quite smoothly.

After our final run and my first simulated crop dusting venture, I flew the Stearman back to the main airport. We both had previously agreed to meet at the airport office after the flight to make plans for tomorrow.

At the end of our meeting, Dick said, "Mark, I've finally got all this dumb paper work completed. I have filed it with the local CAA [Civil Aeronautics Authority]. "It's official now, all of the waivers are in your name."

Hmmm, now that's odd? I thought. Why my name? It didn't take long, I soon found out after I shot a few questions at Dick. It seemed as though he currently only held a Private Pilot's license, and was not legal to do any commercial flying. We were getting paid for Crop-dusting and this in the eyes of the CAA, constituted commercial flying.

"Tomorrow we start for real! There are a lot of Spittle Bugs that need killing. See you about four AM sharp tomorrow morning, OK? By the way, make sure you hit the sack early. I want you to get started flying right at sun-up," were Dick's parting words.

God! Four AM is an early time to get up out of bed. But, I knew we needed to do this because of the calm wind conditions that generally prevailed during that particular time of the day.

We met the next morning at the Kent State Airport as planned. Dick said, "I'll load you up here at the main airport with the dust [BHC-Benzine Hexachloride]. I've scheduled you to dust this farmer's 60 acre field right here," he said, pointing to the exact location on the map.

"OK, Dick, I'll wait here at the main airport about a half an hour, before I take-off and meet you at the farm."

"Fine," he replied, as he departed in his truck for the selected site. The sun was just now starting to break on the horizon as I fired up the big 220 h.p. radial engine.

After it warmed up, I swung the ship on to the runway and took-off for my first job.

Man, oh man! I said to myself, this damn ship flies a bit sluggish, as I struggled for altitude. I wonder? Just how much dust did he actually put in the hopper? It felt like I had over a ton of dust. Naw, it can't be. This hopper only held about 500 pounds at the most.

It only took me a few minutes to reach the farm and as I circled over the fields I looked for Dick and his truck.

Sure enough, there he was at the far end of this one field frantically waving his arms at me. So far, so good. This farm had several large flat fields with no obstructions.

Dick pointed out the specific field I had to dust with his flag, and I began my run for my first crop dusting job.

I think good, no wind. I can reverse my course and not worry about too much drifting of the dust.

I dove down, opened the hopper, and WOW! The dust came out in a huge cloud. The lower wings helped roll the dust out in a big swirl and from my position, it appeared as though the dust covered everything, including Dick and his truck. I climbed up and reversed my turn, I could see my exact path clearly outlined by the dust I had just dropped.

I lined up for the next run. Everything went perfectly, as I found myself flying through my own dust trail I had just laid down. A couple of more passes and I'm finished.

Well now, that wasn't too tough, I tell myself, and to think I'm getting paid to do this.

Back at the airport, Dick was quite pleased and said, "See

you later this afternoon, we have another job lined up for early this evening."

I got home about 8 AM and expected my wife to greet her warrior pilot with open arms. Instead, all I got was, "UGH! you stink!" (That BHC did have a rather peculiar smell.)

"There is no way you are coming into my house with those clothes on, and you had better take a shower, that smell makes me half-sick."

"Could be that baby you are expecting," I said to her.

"No Way," she said, quite irritated.

Later that afternoon and evening, everything went great. I arrived home shortly after dark. As I greeted Ruth, I made the same impression as I did in the morning, even though I had changed clothes at the airport after completing this job.

Greeting her at the door (she still wouldn't let me in), she said, "Mark, I feel the money is not worth the risk. You must end this foolishness right now, and quit."

Pleading my case I said, "But Ruth, I'm very happy. I'm back flying again."

"Why, Mark, you haven't been paid a single cent, and you are risking your life and our family's future," she moaned.

I got a phone call from Dick that night saying that he had another job lined up for tomorrow.

Ruth was not pleased by this at all. She felt that my school work would suffer for sure. "There was always the possibility of you getting killed doing this stupid thing."

Brushing aside her fears, I said, "Please dear, you're worrying for nothing."

I met Dick early that next morning and he greeted me saying, "Hi, Mark, I flew the ship over to the farmer's field last night so we could get an earlier start. It's all loaded and raring to go."

It was still dark as I made my way over to the ship. "You know, Dick? Something has to be done to fix up this hopper door. The chemicals and the metal are reacting with each other. The aluminum hopper and the steel door do not make for a good combination. It's hardly working, I can barely move the door and it creates a big distraction for me while I'm flying."

"Yeah, I know Mark. I'll be sure and take care of that later," he replied. "Oh, by the way, I didn't get a chance to check the field. It's a tight one today. So be extra careful," he cautioned me. "I'll be out flagging you on each and every run."

I really struggled to get this ship off the ground this time. Damn, "Dick must have put more dust in there than he should have," I muttered out loud. I had no doubt that he was most anxious to get the job done in order that he could get his money.

Can't say as I can blame him, I only hope that he does. I would also like to get paid as I feel it would take some of the pressures off me. Ruth has yet to see a dime from all my efforts and this has not helped matters one bit.

Usually either Dick or I always personally go out and walk each field prior to flying. On this particular morning I honestly felt I was being pushed by Dick as he had moved the ship, and was anxious for me to get flying right away.

Dick, noting my concern said, "Don't worry Mark, you won't have any trouble."

As I made the first pass in this early dawn light, I was just able to locate Dick waving his flag on the far end of the field.

The early morning fog and low visiblities made it rather difficult to see everything clearly. I made my pull-up and turn-around as usual, and began my second pass. I dove down on the field and flew over Dick once again thinking, I can't for the life of me, see how he can stand staying in all the BHC dust. He gets quite a thorough dusting on each and every pass.

Now, he is waving at me to indicate that I should get down lower. Lower! . . . Lower!

"For God's sake, Dick. I'm almost on the deck now," I screamed back at him from the top of my lungs, knowing full well he cannot hear me over the roar of the airplane's engine.

As I started my third pass, he waved me down. Down! Lower yet!

I pushed the control stick forward to comply. Suddenly, I felt a sharp jolt, as the ship came to an almost abrupt stop . . . I screamed, *"OH NO! WHAT IN THE HELL DID I HIT?"* knowing the only one who could hear me was myself.

The ship was vibrating and shaking badly. I thought for

sure I had lost one of the big propellor blades that powers the Continental engine. I had no choice. I had to land just as quickly as I could, before the engine shook itself off the engine mounts.

After I successfully landed, I killed the engine with the mixture control and quickly jumped out on the wing and on to the ground. I closely inspected the ship in an effort to try and determine what had caused all this vibration.

It didn't take long for me to see exactly what had happened. The small agitator propellor located under the left wing of the airplane had lost one of its blades.

This little propellor is used to stir up the dust in the hopper, which in turn helps keep the dust flowing out of the hopper smoothly. This was the cause for all the tremendous vibrations and shaking that I had felt while I was flying.

Further investigation revealed I had over 15 feet of electric wire wrapped around my left main wheel and landing gear. It trailed from the gear back beyond the tail of my ship. I silently prayed, thanking God that I was able to get down safely. I'm damn lucky to have gotten down alive in one piece. It wasn't too long before we had the local sheriff visit us.

He understood our problems and his only comment was, "Glad you didn't get hurt. By the way guys, a lot of the local farmers are quite irate. Their power is out and they can't milk. There's also a hatchery with its incubators and thousands of chicks that could be in deep trouble if they don't get the juice back on, and soon."

Turning to Dick, I said, "Oh Lord, this could cost a lot of money, Dick."

Naturally, Dick didn't think too well of me after that little statement.

Returning home, I told Ruth of my little episode. "I know," she said with an icy tone and disgusted voice, "that little fiasco just happened to be on the radio, and I heard all about it."

"But . . ., but . . ." I was cut short, when Ruth gave me this stern lecture. "You could have gotten yourself killed by doing this stupid, dumb-ass thing, you call crop dusting."

Her final ultimatum was, *"Either get out of this business, or get out of my life!"*

I got out of the crop dusting business.

This crop duster/sprayer pictured here is busy at work spraying the crops. Photo by Craig Smith: Courtesy of SRA (Stearman Restorers Association).

Chapter 2
"It's Either . . . A or F"

Those exact words were spoken to me by my Professor at the close of the last quarter of my senior year in college at Kent State University. How did I ever get myself in this dilemma? As a senior, we did enjoy a few privileges that were unknown during our early years in college.

I had originally enrolled in the school of business at KSU primarily because my girlfriend in New York felt I had to go to school for a degree in business, if I ever had any hopes to succeed in this world.

It was tough going getting out of the Air Force, now back in school, five days a week, with about 10 hours of studying and no more flying.

My mind was hard-pressed to accept these new restrictions. Mid-way in my first year, I received a special letter from the Dean of the Men's Business College. *"You are now on probation. You must get your grades up this quarter, or you will be dismissed from school."*

What a blow! For one thing I really hated this school, this course, and this state. The weather was horrible. It started raining in the early fall, turned to snow later, and in general, was just miserable. My girl friend in New York, quit writing and to make matters worse, we had very few girls in our college. The

G.I.'s had flooded the campus and the boy/girl ratio was poor by any odds. 20 guys to every girl.

Things looked bleak for me, and I had almost convinced myself it was time to hang it up and go back home to New York.

One day before class started, a situation developed that would dramatically change my life. A nice looking blonde accidentally (?) bumped into me, and both our books went all over the hall. I asked her her name and invited her out for a cup of hot chocolate. She was nice, but politely refused my offer.

The next day in that same English class, I managed to sit next to her and found out that she was an ex-Marine, a G.I. I tried to get a date and asked her again if she would like to go out with me.

Her reply was quite emphatic. "I don't drink milk, I was raised on a dairy farm, and I hate milk," she replied.

"OK, then, how about coffee?" I asked. Now, that was a different story.

Well, that's how it all began. My life changed very much with her help, understanding and love. I stayed in Ohio and school. I changed my Major from Business to a Double Major in Aeronautical Technology and Industrial Technology, with a Psychology Minor.

I made a new application for a change of colleges. It was extremely tough getting permission. The Dean of the new college, looking over my past college records, frowned on my entrance and didn't think I stood a ghost of a chance since my grades were so low.

But, with Ruth's prodding I persisted. I was given a one quarter probation period by the Dean to see if I could possibly make it. As a starter I carried 18 hours. Where as before, I had only carried the minimum of 12 hours.

I went forth with a new found purpose in my life. For the very first time, I felt I knew where I was going in this college education program. (I later graduated with a 3.2 average.)

One of the more interesting courses was conducted by a man named Mr. Oliver Luey. Most of the ex-GI's got along real well with him.

"Savage," said Mr. Luey, "You seem like the type of guy

who likes a challenge. Are you ready to accept this one?"

I had no idea what he was referring to, but I decided, what the hell, why not. "Sure, Mr. Luey, I'm game. What do you have in mind?"

With this kind of acceptance, I was presented with this option to taking the final exam. I was assigned a project that had to be completed by the end of the current quarter. This particular program was a course in Aircraft Engine Power Plants.

Our class was made up mainly of ex-servicemen from every branch of the military. Mr. Luey was an excellent teacher and gave everyone of us an opportunity to do a lot of projects that would help us later on in our chosen field of aviation.

I was in my senior year and eagerly looked forward to working and learning more about aircraft power plants.

Noting my interest and enthusiasm, Mr. Luey enquired about my background. His particular teaching method was to seek out the student and try to guide him towards the field of his interests in aviation. We had both hit it off right from the beginning of this course.

It looked like for a change, this course would not only be fun, but would turn out to be a breeze course as well.

After the first two weeks of the quarter, Mr. Luey approached me and said, "Remember my asking you about accepting a challenge? Well, with your aviation background, I would like to offer you, as a senior, a real personal challenge."

Not knowing what to expect, I thought, Well, why not?

"OK, Mr. Luey, I'm game. Tell me what you want done?"

"See that aircraft engine mounted on the run-up stand?" he said pointing in the direction of the small engine mounted on an aircraft run-up stand.

"Yes sir," I replied.

"OK, I want you to take it completely apart. Using the shop's Micrometer, mike all the parts. Get new parts if needed, set up a parts list manual and then reassemble this engine."

"Great," I replied "as long as you feel that I will have enough time."

"Oh, by the way," he added, "your final grade will be given when it runs."

"What do you mean by that?" I asked.

"If it runs, you get an A. If it doesn't, then you will get an F. It's, either-or," were his parting words.

Now this was right up my alley I thought, as I breathed a big sigh of relief. This was a very critical point for me, as I was just about ready to graduate and I couldn't afford an F.

I worked diligently on this 65 H.P., Continental engine during the quarter and was beginning to wonder whether this particular project just might be a bit bigger job than I could handle. Time was running out, as the big day for the final exams was rapidly drawing closer. The final exam day had arrived, and as of this time, I still hadn't gotten this damn little engine to run.

The rest of my classmates were in the big glass covered classroom deeply involved in taking the finals, as I diligently labored out in the main shop, desperately trying to finish up this engine and get it ready to run.

Mr. Luey sensing my frustrated feelings walked out of the classroom and said, "I can't make any exceptions, Savage."

"I understand," I replied, and went back to work on the engine, wondering if I had made the right kind of a deal.

I finally had this engine assembled and prepared for the big moment of truth. It was time now to get this little engine to run. I was beginning to feel a bit of panic starting to set in. This little engine would not start.

Even after all the cranking, I couldn't get it to pop. The class period was almost over, when Mr. Luey came back out again to watch. He wasn't alone.

I had every eye in the class centered on me, despite the fact they were taking their final written exam, they took time and watched me like a hawk to see how my project was doing. They all knew what was on the line.

Mr. Luey, probably feeling sorry for me said, "Savage, remember? No exceptions."

"I understand," I replied, as I wiped the sweat from my brow.

"Why not try shooting just a bit more prime into this engine," he said, "Only this time put some fuel in each one of the cylinders, and get one of your buddies to help you."

"But, do you think I should put that much fuel in it? For one thing it is not tied down. It's only mounted on this aircraft mounting test stand and it might not hold."

"Oh, I don't think it will hurt. All I need, is to see it fire just once," he replied. "Get your friend to prop it for you."

Yelling for Tom to help me, we added the fuel and tried again. The end of the period was here and all of the guys were looking out of the classroom grinning, and watching me sweat and I hope, feeling sorry for this dumb jerk.

I was at the rear of the engine and on the throttle controls. Tom agreed to prop the engine. It had no starter and required someone to swing the propellor to start the engine.

"Switch on! Contact!" Tom yelled. Then, "Are you ready?"

"As ready as I ever will be." I said, with my fingers crossed and a short prayer.

One pull on the prop, nothing. Next pull, still nothing.

This looked like the end, and an "F" will be my final grade.

"Let's try it one more time," I pleaded.

"It's useless," Tom replied, "I don't think it is going to run."

"But, why not Tom, what have I got to lose?" I pleaded.

I noted how much Tom was sweating from all the work, yet not as much as I, and I wasn't even working.

Once again, the same familiar routine. *"Switch on?"* asked Tom.

"Switch on," I replied.

"Gas on?"

"Gas selector switch is on," I replied.

"Here goes nothing," said Tom.

"Contact," yelled Tom, loud enough to get everyone's attention.

"Contact," I replied, hoping against hope this damn thing would pop, just once, to save me.

Then almost by magic BAM! . . . BAM! . . . BAM! . . . the little engine sprung to life with a roar. I tried in vain to reach for the throttle as the test stand took-off in the direction of the large amphibian Goodyear "DUCK" airplane parked in the center of the hangar/class room.

I'm still hanging on to the stand with the engine roaring as it took off in its uncontrolled flight. It was at this point we both parted company.

The stand with its Continental engine roaring, suddenly and dramatically swerved away from the Duck and past a partially assembled Piper Cub. The noise was deafening as the little engine with all its unbounded energy, went crashing through the metal and glass partition wall that separated the classroom from the shop area.

The students taking their written exams in the classroom saw what was about to happen and vacated the classroom in one hell-of-a-hurry.

Pieces of glass, metal, and propellor went helter skelter in all directions, as the little engine with its stuck throttle at full power setting, was rapidly destroying the almost brand new classroom.

Two, now three, table/chair desks fell victim to this wild raging monster. Pieces of the prop were hurled up and lodged in the thirty foot high ceiling. Finally it was over. All was deadly quiet. So very quiet, you could actually hear a pin drop.

No one knew exactly what to say. I estimated that several hundreds of dollars damage was done to this area. I tried to be calm as possible but my legs were still shaking and like rubber.

Mr. Luey stood silently in the room. His mouth agape, as he shook his head in utter disbelief. He very slowly walked over towards me. Before he could say a word, I said: "Well, Mr. Luey, it ran didn't it? Now, do I get my A?"

He was absolutely speechless, and could only nod his head.

Yes . . . I really did get an A+ in this course.

Chapter 3
Pennies from Heaven

Back in the early fifties when I was a representative with the State of Ohio and the Ohio Aviation Board, I felt that I could help in the organization and promotion of glider clubs to try and encourage the young men and women in colleges to get into aviation.

Actually, I became interested in gliders in WW II, while serving in the Air Force as a fighter pilot in Italy. I learned later that the German pilots had acquired their skills starting as young glider pilots.

One other time I had slight contact with the gliders while living in Spokane, Washington. I spent many an hour on my back watching the gliders circle overhead at very high altitudes. I always wanted to try this sport, but never did seem to take the time nor was I able to get any information about them.

I had no knowledge of gliders or sailplanes, and felt that the only way to proceed would be to visit other clubs in and around Ohio, scan their operations for ideas, and then mould or pattern a club around them.

My first ride in a glider was at Dayton, Ohio with the SSD (Soaring Society of Dayton Club). I took my first aero tow and flight with a guy by the name of Rudy Opetz. He took me up in a Schweizer glider. Rudy just happened to be one of those German pilots I referred to earlier.

He had been a well-known test pilot with the Luftwaffe during WW II. He had actually tested and flown one of the very first German twin jets during WW II, and was well known in the field of aviation in the U.S.

Next, I visited a Ft. Wayne, Indiana club and their auto tow operation. They had a Schweizer single seat glider. This club was small, and operated out of Smith Field near the northern edge of Ft. Wayne, Indiana.

The two characters I met were versatile, and friendly. Both Tennis "Moe" Mahoney, and Harley Schaffer, made me feel like we were old friends. They inaugurated me into launching a glider via the "auto-tow" method.

Their club had a 1937 Ford Tudor Coupe for ground towing. Harley drove and simply said to me, "Hang on," while Moe flew the glider on the launches. Harley began the ground launch with a wave of his arm. Moe in turn waved back. And, away we went in a cloud of blue smoke. The Ford burned a little oil, and the initial start always produced a huge cloud of blue smoke.

As it turned out, Harley's advice was an understatement. My fingers literally dug deep into the seat, as I clung and hung on for my very life. Wild and Wooly, was the name of this game. Yet, the bird did get airborne and seemed to get a fair amount of altitude for this type of operation. Just as the Ford approached the very end of the airport and fence, Moe released the rope from the glider.

Since Harley and I were much too close to the fence to brake and stop, Harley simply turned the steering wheel of the car in one direction just as hard as he could.

We wound up doing a beautiful ground loop with the Ford. This in turn always created a huge cloud of dust that could be seen for miles away. My heart was beating very rapidly as we came to a stop in this man made cloud of dust. Harley secretly admitted to me that the car's brakes were a bit weak.

We spent the rest of the afternoon together, as they both demonstrated for me the virtues of a club using the auto tow as a method to launch a glider. Both Tennis and Harley went to

The winch in operation at the Ohio State University Airport. The winch was built by OSU students of the Buckeye Glider Club. Jim Greshel in the spotted shirt on the left and Ralph Turner waving his arm.

great lengths to prove that it wasn't as bad as it first seemed.

I thanked them both, and felt that I had been thoroughly schooled in the art of towing a glider by using the auto as a launching system.

Next, I went to Detroit, Michigan, and watched the smooth operation of the "Vultures" club, using a winch. They did a very good job and managed to show me at least a dozen well launched flights that went without a hitch.

Still not completely satisfied, I decided to visit the home plant of Schweizer Aircraft Co. They were building American gliders and sailplanes and were located in Elmira, New York. This place was also known as the sailplane capital of the U.S. It was here I met the three Schweizer brothers. At their suggestion, I was advised to set up a meeting at the Aviation Board in Columbus, Ohio.

I invited all the men who were knowledgeable in soaring to my office in the State of Ohio Aviation Board for a meeting. My initial plan was to get as many professional men who were involved in this sport of soaring to come and express their ideas.

I felt it was important enough that I should know how I could go about setting up a glider club.

Upon my return to Columbus and the Aviation Board, I sent out special invitational letters to all soaring interests located throughout the U.S. I was very pleased when twenty prominent leaders in the soaring industry showed up.

We all had a very pleasant meeting and lunch, until I brought up the suggestion that perhaps now would be the time to state their opinions as to whether a club should start operations on a winch, auto, pulley, or airplane tow.

At this time everything came apart at the seams. It seemed as though every one of my guests had taken a very firm stance as to their feelings about the particular type of tow method that they would support, and, there was absolutely no compromise.

It didn't take long before I became exhausted and totally disgusted trying to moderate this discussion, when old "Doc" Raspeth, a well known aeronautical engineer from Mississippi State College, came forth and solved my dilemma.

"Mark, why not form the club and do it the way you

want?" he said.

This was the innocuous beginning of The Ohio State Buckeye Club.

Now, armed with this encouragement from Dr. Raspeth, I began to feel that my way was clear to proceed in the formation of a bonafide college glider club. I posted a notice on the Student Union bulletin board of the Ohio State University.

NOTICE!

Anyone interested in flying gliders/sailplanes come to the Glider Club meeting to be held next week in the OSU Student Union Building at 8 PM sharp.

This initial notice produced a SRO (standing room only) crowd of 300 plus students. All eager, and rarin' to join.

The first half hour was spent in discussing gliding and soaring and sailplanes, plus all the fun it would be flying them. The inevitable question finally arose. "Sir, just what kind of equipment does this new OSU club have?"

"Ah . . . ." I stammered. "We really don't have any equipment at this time."

After this, I was forced to confess that this new group was to be the nucleus of the club. My plan was to solicit money from all the students who would volunteer and sign up.

With these cold facts in hand, the SRO crowd melted rather quickly, leaving me with a small group of students. A die-hard cadre of ten kids who got together with me to form the Buckeye Glider Club, of the Ohio State University.

By-Laws, Constitution and operation offices, all had to be set up and acted on. It was a long cold late winter night when I arrived home dead tired, but I was flushed with success. I then decided that maybe I could promote some type of an automobile to be used in building a winch.

Inquiry was made at the Columbus Police lot and for $10, we found a 1948 Buick car with an automatic transmission.

We also needed a field car. This task fell on my shoulders to try and locate one for free, or next to nothing cost. I was reminded again by my new group that the club had very little money.

As luck would have it, I found a former Air Force type guy who was sympathetic to our cause. He just happened to be a used car dealer, and agreed to give us a 1936 Ford Tudor that we could cut up and make into a field car. In turn for this gift, I promised him that we would get him all kinds of publicity when we were operational.

I divided the students into two teams as we began the long hours and days of work.

First planning, then cutting, forming and welding at the OSU Rocket Lab. Fortunately, a good friend of mine, Professor Von Eschen the lab boss, was very sympathetic to our cause. Before long, the field car was finished about the same time the other group finished building the winch. These two projects kept me very busy.

The club had an advisor and he felt that it was his duty to work with the students and guide them through this growing-up stage. He did have experience in gliders, so the kids looked to him for help.

Following his suggestions, the club bought 5,000 feet of piano wire to be used on the winch. With the winch finished, it was time to test out the field car at the Ohio State University Airport, better known as Don Scott Field.

One day after making extensive repairs on the car, their advisor, Mr. Fred Tietzel decided that he wanted to test-hop the car on the airport. He got in the driver's seat and immediately began to race like hell down the field for all the car was worth. The three of us in the car had to hang on for dear life. POW!! . . . POW! . . .! in and out of that pot hole like a shot. Well, the car was still running. Another bump. . . . Whew! Then a tremendous jolt.

The car started belching like a steam engine and we barely chugged back to the anxious college kids waiting for us.

A hose had almost been completely torn off the radiator, and the car was wheezing and acting like a tired broken down old work horse. I was fit to be tied and just as hot as the car.

The students were not about to buck their class advisor. But I could, and did. He was told by me in no uncertain terms to get his butt away from this club.

The field car discussed in this chapter was built from a Tudor Ford donated by a local Ford dealer. Driver unknown, with Jim Greshel seated in the rear.

It was from this point on the club began to operate quite smoothly and harmoniously, with a new club advisor.

The winch was built, the tow car finished and the club ready for flying. But, where was the glider?

The kids looked to me to find the money and ship for them. It was indeed a tall order. I had been working for the State of Ohio, and it was a bit tough to go out and try to promote money in this position.

My particular job with the State of Ohio Aviation Board was to inspect all of the airports in Ohio. Our department on special occasions had to fly some of the key people around the State for meetings.

I had been flying John W. Brown, the Lt. Governor of Ohio, around Ohio to several meetings. We had become pretty good friends. Here is where I would find out just how strong our friendship really was.

One day after a very pleasant flight and a successful dinner meeting, I hit the Governor up with my proposition.

"Sir, I want to tell you all about my plans for getting a glider and glider club started with the kids at the OSU University."

"I also plan to get glider clubs started at all the other State Universities in Ohio. BUT! I need your help in getting the first one going."

"Oh? is that right, Savage? And, just how do I fit into this picture?"

He was awed, and utterly flabbergasted, at my pitch. Then he said, "OK, sounds great, Mark, but just how can I be of any help as the Lt. Governor?"

This was exactly the point I needed, as I said to him, "Well, Sir, all I need is a mere $1,000 from you to buy a glider."

"You need WHAT? . . . For WHAT?" he exclaimed, just a bit stunned.

"It's for a good cause, and I know that they will be able to pay you back . . . someday," I continued.

"Well, now, if that don't take the cake," he replied. "I have had a lot of propositions thrown at me in my time, but I think this one is the most unique," said the Governor.

Not wanting to get him upset, I held my breath, not knowing just would happen next.

"This is beyond a doubt the craziest thing I have ever heard of, Mark. But, I am convinced that with your help, the club will succeed," replied the Governor.

That night, I had a check from the Lt. Governor for $1,000. It was like a gift from heaven. The OSU club members paid him back in full, several years later.

A few years later, the graduate members of OSU decided since they were no longer eligible to fly in the Buckeye college club, they would form a new club to be known as CSA or Columbus Soaring Assoc.

When the Buckeye glider Club and CSA joined together a few years down the road, a new club was born. This in turn became known as, the Central Ohio Soaring Association, Inc. or COSA.

As a point of interest and comparison, COSA now has 142 members with a total of 24 sailplanes. 2-two ships, 2-trikes, 1-Ten plane "T" hangar, a 100 ft. x 120 ft. quonset type hanger and, one very used postal van.

Chapter 4
Up! Up! And Away

The Columbus Soaring Association, better known as CSA, had been operational on just about every field in and around the Columbus area. The club itself could not call any particular place home. Needless to say, every time we moved to another airport, we usually had at least one minor accident within a week or two after we started operations.

It wasn't that the guys were dangerous, it's just that every time we changed sites, we took time getting used to the new areas and we did make mistakes. The club had experienced members, but it also had acquired a higher performance sailplane that was able to soar much better than the Schweizer ships most of the members had trained on.

This ship was also a bit skittish. It took a lot of getting used to, by everyone who flew it. The Cinema two-place sailplane was bought from a club in Toledo. It was our thought we would develop CSA along the same lines as the Buckeye Glider Club. The nucleus of the club came from the alumni of the Ohio State University glider club.

We had to go through the same process of buying a car, then converting it to a usable winch. Only this time it was different. All of the charter members had cut their teeth on the building of the original winch. This new winch was built in jig time.

I felt that this was the natural way to go in order to develop this new club on a businessman's type of operation and the growth should come from the people who would eventually own their own sailplanes.

A Buick convertible was located in the police impounding lot. Through the help of Turner and Marsh, another winch was built, only this time it would be far different than the first. It was designed to be portable. We mounted the wire drum on the right rear wheel and set up a guide channel alongside of the car on the right fender.

The idea was that if the unit had to be moved to the other end of the runway, our only requirement would be to take off the drum with the wire on it and then remount the wheel. We could then drive the winch off to the other end of the field, take the wheel off once more, remount the wheel with the wire and we would be back in business.

In theory, it was a masterpiece, but in actual practice it was a colossal flop. No one wanted to take the time for the conversion. The drum and the wire on it proved to be just too heavy to lift on and off easily.

Our operations began at a small airport located just south west of Columbus, Ohio. At that time there were only six airplanes based there.

The bulk of that fleet was owned by a friend of mine who operated a crop dusting service.

From that humble start, we managed to operate several years at this airport even though we did try out several other airports in and around the Columbus area.

Unfortunately, most of them did not work out for our operation with our Cinema sailplane, and the Buick auto/winch combination.

Our group made a sad picture with its equipment. We were quite a rag-tag bunch of businessmen, compared to the now flourishing Buckeye Glider Club.

For different reasons, CSA was forced to operate from various fields. The combination of equipment, experience and airports set off a rash of minor accidents at just about every location. These incidents were not serious but downright disgusting. The Cinema usually required a minimum of at least twenty tows to get a member ready for solo.

The Cinema was a bear for floating. Slips were an absolute must on every landing. It was easy to spin and it had a built in ground loop on every landing. But . . . it could soar well, after a little knowledge and practice.

One particular week-end, while operating at an airport located east of town, Turner got the bright idea of putting two extra drag chutes on the wire. "In that way," said Ralph "we can increase the drag on the wire and be able to get it closer to the winch at the end of the tow, and it sure would save us all a lot of retrieve time."

The theory was that the wire would fall straight down, rather than coiling, after its release from the sailplane.

"This idea has real merit, and I agree with you. It should work fine, more drag on the wire after release, so therefore the wire should come straight down with no effort," I said to Ralph to encourage him.

Fortunately, as it turned out, I took the first tow and went solo. It was a beautiful winch tow by Bob Miles on the Buick Winch.

I was able to reach 1500 feet rather quickly without a bit of trouble. A good tow, the right height, I said to myself as I got ready to get off the winch tow.

Reaching for the red knob release, I pulled and waited for the loud bang that signifying I was off the wire. Strange, NOTHING! . . .

Did I really hear right? NOTHING! I pulled the release knob again, only this time much harder. Still nothing! Frantically, I pulled the release knob several more times. But, I'm still on the wire! My heart began to beat a bit faster and I feared the worse. Now the winch was visible over my right shoulder. As I looked down, I noticed with alarm, I've over-run the winch by at least one thousand feet. I felt my only chance to survive would be to make a diving right turn back to the winch, and try to land down-wind. Hopefully, the wire will detach or break, before I am snapped up on my nose and over on my back.

Horrified, I see Bob Miles bail off the winch on a dead run as I dove for the winch. I could hardly blame him. All he can see is the Cinema aimed straight down on him with all the trailing wire about to coil around him and the winch.

The wire was everywhere. Three thousand feet of a coiling, hissing wire. It was like a bare fanged viper, poised and ready to cut its victim up into little pieces.

Suddenly, a sharp jolt and I am free. It's true, but I can hardly believe I am so lucky to be free.

I think, is this for real? Putting the ship into a big slip, I successfully managed to land and stop on the last quarter of the runway . . . downwind!

"Oh what a relief!" I said aloud as I breathed this big sigh.

But, what about Bob?

A car finally picked me up and we drove back to the now empty winch. It was an incredible sight. Bird-nests-of-wire were everywhere. . . . Trees, barn, house, electric light wires, and all over the winch. Fortunately the farmer saw what was happening and rushed over to cut off the electricity.

It took hours and hours to cut through the maze of tangled wire. Needless to say, this act shut down operations for that week-end.

"Well, Ralph," I said. "It looks like that was one hell-of-a-fiasco."

"No kidding, Mark, but we are all glad you didn't get hurt," replied Ralph.

That next weekend we were back in business. This time we stuck to our proven single drag chute operation, rather than the multiple chutes that we had tried earlier.

We were very busy making our usual launches one week-end and were completely unprepared nor did we anticipate what would happen next.

As it so often happens, one of our new students, Harry McIntyre was on this particular winch tow.

It was a good tow, and he did quite well, until he tried to slip. He was much too high and held the slip too long trying to lose altitude in order to land on the field.

As he lined up with the runway he held the slip too long and this resulted in a loud ker-plop. The tail actually came around and met the nose. Our first causal visual examination convinced us that we would have a major repair on our hands. What to do?

Another one hundred dollar assessment had to be made to keep this businessman's club alive. With this new money two other members and I volunteered to help repair the ship.

The annual Wright Memorial Glider Meet was drawing close and we particularly wanted to go there and show off our ship.

The interest in the club was quickly waning. We had to act, and act fast, in order to hold the group together.

The Cinema was taken to the Ohio Aviation Board's brand new hangar. The cement on the new floor had hardly dried. In fact, the Ohio Aviation Board's new office and hangar had not even been used or officially dedicated yet.

This did not stop us, we had to get that Cinema sailplane finished in time for the annual meet that was coming up during the Memorial day week-end in Dayton, Ohio.

We spent a lot of hours working nights just doing the required large welding repairs.

We were lucky and completed this welding in about three nights. We spent quite a bit of time and effort in the hangar in order to meet our set schedule.

Our wives thought we were all nuts. None of us ever made it home much before midnight every night. They were concerned that we would be unable to work the next day because we had been working so hard most of the night.

With the welding and repairs completed, the frame was soon prepared for the final covering. Fabric, dope, and then paint.

As the time came for painting our bird, the weather started turning bad. It had been raining on and off for the past week. The forecast called for it to continue. Our head man in charge of maintenance Larry Chalfant, had been our leader. It just so happened that he was the only one at this time who had an A&P mechanic's license.

"Say, Larry, do you think we have a ghost of a chance to finish this ship before the scheduled Wright Memorial Glider Meet next week-end?" I asked.

"It's going to be tough, if we don't get a dry day soon," he replied.

Towards the end of the week we were ready to apply the first coat of dope. Even though it was still raining, this did not stop us.

Paint brushes in hand, we slapped on the dope. "Do you think it will dry enough by tomorrow night Larry?" asked "Mac".

"Don't worry Mac, I think we will be OK by then," replied Larry.

Another night and another coat of dope. We seemed to be making good time. Fortunately, the weather did cooperate a little bit.

This in turn helped us finish up the doping on the fuselage. The closer we got to finishing up the project, the later we worked.

Now it was close to two in the morning by the time we hit the sack. Oh well, all we have left to do is apply the silver and then the color. It will be close, but it looks like we just might make our deadline.

We were rapidly closing in on the last day of the week and the Wright meet. Later that week, Larry brought over his portable spray outfit and we soon were busy applying the silver dope to the fuselage.

The next night Larry said, "It's time for the paint."

We all decided to use a Kelly green paint for the fuselage.

Unfortunately, it was raining again, and even though Larry was a competent mechanic, I was deeply concerned about all the rain and the heavy downpours that were soaking the entire area and causing a lot of heavy humidity.

This prompted me to ask, "Do you think we can paint in this heavy rain and humidity, Larry?

"It will be tough Mark, very tough. But, I'm sure we can manage," replied Larry adding, "Yeah, it will be hard, and there's no question the humidity is as high as it could be, with all this rain."

I thought for sure he would call off the painting for this evening, particularly with the recent remarks he had just made. Not so.

He loaded up his spray gun and was ready for action. He whipped the spray gun back and forth in a feverish effort to get the paint job completed for our deadline. The spray gun dripped water constantly and so did the rain. Larry kept adding retarder to prevent the paint from blushing.

It wasn't too long before the Kelly green paint christened not only the ship, but the hangar floor area as well. We had to use every trick in the book in order to keep from ruining the job.

The Cinema was finished in time for the annual meet, and I said to the guys with me, "I think I will be finished too." Paint, papers, tape, etc., were everywhere.

"Oh, what a horrible mess in this beautiful new hangar, and it hasn't even been dedicated yet!" I moaned.

"Don't worry Mark, we will help you clean it all up," said Larry as he grabbed the fire hose and began washing down the floor.

The meet day arrived and we did manage to get the Cinema over to Dayton for the annual Wright Memorial Glider Meet. As it turned out, it was the hit of the show.

After the last day of the meet, it took a few more nights and a lot of hard work but we finally got the hangar all cleaned up. Why, even to this day, there is a smidgen of Kelly green paint still on the floor of the Ohio State hangar!

Chapter 5
The Air Show

Both Ralph Turner and Wally Marsh, former members of the Buckeye Club, were now working at the North American Aviation plant in Columbus, Ohio. Over the past few months, both had helped me recruit a whole new group of people from the NAA Company. The winch was sold and we searched around and found a relatively low priced Waco UPF-7 that this new club known as CSA decided to use for a tug. With this new tow ship and the Cinema 2 place sailplane paid for, we were all ready to fly. It was by far so much easier this time as a lot of the guys were working full time. Money was not as tight as compared to the early days of the college club.

The Waco airplane was called a UPF-7 and its beautiful new paint job gleamed in the sun. It had been a military trainer during the war and this ship could take a beating and still go.

Unfortunately, it was a lot more airplane than most of the members had ever flown. Fortunately, several of the club's members were ex-military, and it was a piece of cake for them. Even though the CSA club was growing, there just was not enough new people getting involved.

CSA had a total of 20 people and five gliders. This was by far much bigger than the Buckeye Club. Basically, it was

The Waco UPF-7 tug is shown towing the Schweizer 1-26 glider. The large crowd on the ground got this same view, except at that time the Waco was towing the Cinema two seat sailplane.

formed along the same lines as the other club.

We had found in the past that in order to survive and fly at a reasonable rate, the club constantly needed new members.

One week-end during the winter, Wally Marsh, McIntyre and myself decided it would be nice to go down to Chillicothe and make a few flights to get current. After all, winter had set in and it was almost two months since we last flew.

It was very cold that day as we rolled our equipment out of the hangar. A winter storm had passed through the night before, and the ground was covered with about a four inch blanket of snow.

I had agreed to let Wally fly my sailplane. It was a small single seat Breigleb BG-7. Mac would be flying the club's Cinema two place glider. Me? Since I was the only tow pilot, the towing fell on my shoulders.

The weather had cleared and the day dawned bright and beautiful. The remnants of the cold front left a few cu's (cumulus) that began to pop out in the sky. Both Wally and Mac were anxious to get into the sailplanes and go. Just because we had about four inches of snow all over the runway didn't mean they could not fly.

Mac won the toss for first tow, and I towed him off first. The Waco's prop make such a huge cloud of snow behind me that it completely covered Mac and the sailplane on take-off.

The minute Mac's sailplane broke ground, he immediately pulled up and flew just above the big white ball of swirling snow that I had created with the big propellor blades of the Waco's engine. Hmmm, nice flying Mac, I thought.

As I towed him up to the release altitude of 2,000 feet, I noticed that clouds were beginning to develop rather rapidly, so I decided to put him between and above the rather large budding cumulus clouds that had formed around him. I felt he could always find a hole to get through if those clouds got too thick.

After I landed, I hooked up Wally and we took-off in the similar small blizzard that I had created. I could see where my prop wash had created another snow storm that completely covered Wally and my sailplane. He too flew above this and the rest of the tow was uneventful.

As I towed Wally, I spotted Mac, and I started to fly over to him, noting that he was now above the clouds and making large circling left turns.

Wally spotted Mac about this time and released. I dove for the ground and landed. I had gotten quite cold towing these characters and felt that they should be landing within five or ten minutes at the most. This should give me enough time to get a hot cup of coffee at the restaurant located directly across the street from the airport.

I shut down the engine and stiffly climbed out of the cockpit. Boy oh boy, was I ever cold. I wondered how they were able to stand the cold. At least I had a heater in the Waco, but they had nothing.

I returned to the airport in less than ten minutes and felt for sure that both of them would be on the ground by now. How wrong I was. I looked around and was shocked when I did not see either one of them on the ground.

I stayed in my car and kept warm with the heater and wondered, "just how long could those characters stand this cold?" I asked half aloud. "Damn, it's now been over thirty minutes since take-off and I am getting real concerned. There is no way they can possibly stay up." I moaned aloud. "They have to be down somewhere."

Fearing the worst, I climbed back in the Waco to go out and look for them, dreading what I might find. Desperately I looked all over but I can't see them any where. The skies now are almost completely overcast. I looked everywhere, and I'm unable to spot either one. The broken cloud conditions have gone to an almost solid overcast. There are a few holes left, but mighty few.

As the ground was completely white with snow, I felt it should be easy to spot them if they were down, but I had no luck at all. They're nowhere to be seen.

Now I feel that maybe the only other place to look would be above the clouds. "Naw . . . it just can't be," I told myself. Just the same I thought I had better check it out.

I found this small hole and climbed the Waco up on top of

the clouds. I looked everywhere and still I can't seem to see a thing that resembles two gliders.

Then . . . suddenly a flashing light, as the sun reflects on the wing of a ship in a turn.

Holy cow! I just can't believe this. There they are, the two of them, flying together, about three to four miles out on top of the clouds and it looks like they are working a thermal for lift.

Flying over to them, I waved and they waved back. Somewhat relieved, I dove down through a small hole and raced back to the airport.

Almost an hour had elapsed and I am quite concerned. In fact I'm about at the panic stage, when suddenly Wally in my ship appeared first over the airport. He had flown over the field and was on his final approach to land.

After his landing, I ran up to him and all I can see was this frozen smile. He managed to point to his feet and I looked in horror. Both feet are almost completely covered with snow that had been scooped up through the tow hook hole in the nose of the glider.

As I looked into the cockpit I am amazed to see all the snow that had accumulated inside the belly of the ship. I actually had to pry and lift Wally up and out of the cockpit of my sailplane. He could barely walk, but he was still smiling.

"OK, Wally you nut, I'm glad you're down and safe, but . . . Lord . . . I can't believe this . . . Why you only have your house slippers on?"

"Yeah," he said in a stuttering voice, "I didn't think I would be up that long."

Now Mac came whistling in. Both had achieved an incredible one hour plus flight above the clouds.

It was stimulating as well as dangerous. The clouds could have closed in under them at any time.

That next spring in Chillicothe, Ohio, all of our club members agreed that we needed to do something to try and get new members. The local skydivers had contacted us and they too were interested in trying to get new members.

"Why not co-op a program with us?" they asked.

Well, why not? Maybe we each can get a few new members

from some of the potential air show audience.

The sight of our bi-plane, gliders, and sky-divers had diverted a lot of motorists from the main highway to our little airport located north of Chillicothe.

At the briefing, the plan was simple and spectacular. I would fly the Waco with the two jumpers in the front seat. Then Wally would fly the glider with one jumper. We would take-off and climb to 3000 feet above the ground. Then a jumper would leave the glider, followed by my two jumpers. All three planned to exit about the same time. Then, Wally would cut loose his glider and spiral down around the jumpers.

The Waco would circle above everyone and slowly follow them down and would be the last one to land.

"Great plan," said Wally. Each member of the two clubs echoing his words in return.

It was a struggle pulling all that load off the ground and trying to climb up to the agreed altitude. As I circled over the airport, I could see the large amount of cars and people we had attracted to our show.

The old 220 h.p. engine groaned and moaned, but I finally got the Waco, glider, and five grown men with all their gear up to the agreed 3,000 ft. altitude.

My two jumpers got out first, one standing on the left and one on the right wing walk section of the lower wings and looking back at the sailplane we were towing. They were waiting for the signal that would indicate all were ready to make this mass jump together.

I glanced back and couldn't imagine just what the hell was delaying this guy's signal for the jump. He had to leave first as he was the key to the program.

Unknown to me, Wally's jumper was having trouble, he couldn't get out of the back seat. All of his sky-diving equipment had made it extremely tight and he just can't seem to get out of the door and in position to make his jump.

"Get out and jump," yelled Wally.

"I can't, I'm stuck," cried the jumper. Unknown to me, Wally had his hands full flying the sailplane, staying in forma-

OND ANNUAL SOARING MEET to be held at Sky Park Saturday and Sunday drew
crowds of spectators. In this areial picture taken from a towplane flown by Edwi
ner of Washington C. H., can be seen the rows of automobiles lining the airport and
g either side of Route 23. At the top of the picture is Sky Park operations building
he upper right are several of the sailplanes flown at the meet. (Gazette photo b
es E. Leasure Jr.)

A newspaper clip shows the line-up of all the cars and the crowd at the Chillicothe Airport.

tion position and trying to help this guy get out.

I'm still cimbing a little bit as I tried to slow the ship up for the jump. "Just what in the hell is holding them up?" I swear out loud.

Meanwhile, the jumper on my right is yelling, and trying to signal me to, "Slow-Up! . . . Slow-Up!

I tried to yell back at him . . . "Slow up, my ass, I'm almost stalled now!"

He can't hear me over the roar of the engine, and the wind whistling through the wires. Wally's diver is half in and half out of the sailplane. I tried once more to slow up, but the whole ship shuddered like it had a chill.

"Damn . . . I'm about ready to stall. If this happens I'll lose the whole kit and kaboodle," I moaned.

My jumper yells and signals at me again . . . "Slow up!"

Now as he tries to reposition his jump foot, his big shoe missed the wing walk and went crunching through the top of the lower wing.

"Oh, NO! NO!" I moaned, as the ship quivered violently and on the verge of entering a stall and spin. "Well, that does it!" I said, and pushed the stick full forward to pick up flying speed. Both divers are now waving wildly at me to slow up. They desperately try to hang on. The wind is tearing at their bodies and it's all they can do to hang on and not be blown off.

Finally they are forced off, or let go, I don't know which, but they're gone. Looking back I could see Wally's jumper fall free of his sailplane. Wally cut loose from me and banked away to the right.

What a spectacular sight, three parachutists, a sailplane, and myself in a bi-plane circling over them all.

This little performance jammed the traffic on the main Rt. 23 that ran north and south through town.

We must have at least 200 cars parked along the road and airport watching our impromptu air show. The jumpers easily found the airport and landed right next to the crowd.

Wally landed his glider and after I circled for a few more turns I set up the pattern for the airport and landed trailing my tow rope.

As I taxied up to our launch area and cut off the engine, I am met by a tremendous ovation from all the people who had assembled at the airport. I waved my arms in acknowledgment as I climbed out of the Waco.

Wally surrounded by all the jumpers, came up to meet me. "What the hell happened up there, Wally? I thought your man was never gonna jump."

"Well, it was interesting," said Wally. "The guy actually got hung up on the door trying to get out. He was stuck half in and half out. He finally popped loose and that's when your guys saw him leave the ship. Your two guys jumped and as they flashed by me they finally caught up and joined my guy in the mass jump."

"Well that's only the half of it," I said to Wally. "My two jumpers wanted me to slow up to the point where I damn near stalled out and dropped the ship with you on tow. To top all that, the Jump-master managed to put his big foot through the top of the lower wing."

Now, all of the rest of the jumpers came swarming over the Waco and the Cinema clamoring they wanted to go back up and repeat the show.

Both Wally and I responded together saying, there is, "ABSOLUTELY NO WAY, NEVER AGAIN!"

Chapter 6
Tail-In-A-Tent

Wally Marsh had volunteered to drive his car and trailer my glider, a BG-7, from Columbus, Ohio to Richmond, Indiana for the annual WMGM. (The Wright Memorial Glider Meet) that was held during the weekend of Memorial Day.

This generally was the kick off for the soaring season in the Midwest. The meet was called the Wright Memorial because it had originated in the Dayton Ohio area many years prior. The name Wright was chosen in honor of the Wright Brothers, and it remains so even to this day.

The S.S.D. (Soaring Society of Dayton) sponsors this meet and it usually draws soaring buffs from over a five state area. Two weeks prior, I had a mechanic relicense my BG-7 Sailplane in preparation for this big event.

I was personally looking forward to competing in this meet. In fact, both Wally and I were planning to fly part of the meet as a team. I had met Wally and worked with him when he was at Ohio State University after he joined the Buckeye Glider Club. He had long since graduated and was presently an aeronautical engineer at the N.A.A. (North American Aviation) plant.

I was quite confident about my ship being in excellent con-

My Breiglieb BG-7 that provided the tail in the tent and the hilarious night for all of our friends. Shown here after the repair and back flying.

dition since I just had a fresh license. Both Wally and I were eager to compete in the upcoming competition.

Our home based club was known as the business man club and was called CSA (Columbus Soaring Association). CSA had agreed to furnish a tow ship to the Dayton Club for this meet. The tow ship was a surplus military airplane and was called a Waco UPF-7. Over the past year and a half it had proved to be an excellent tug for our club. Just about all of our former launching problems had disappeared when we started using this ship as our tow plane.

Since I was flying the tow ship to Richmond, Indiana for this meet, Wally agreed to take my glider (now on the trailer) over to the meet. It was only an hour's flight over in the Waco and I was anxiously waiting for Wally to show up with my ship. The trip took him better than two and a half hours to drive. Finally he arrived at the Richmond Airport.

I was quite relieved to see him pull up and park next to the other sailplanes. We had used the extra trailer space to bring along all of our camping supplies including our tent.

Everyone pitched in and soon they had the tent up, coffee on and in general were pretty comfy in our portable quarters. I started towing as soon as I got there, and was unable to help with the chores.

Wally, with help, started to take my BG-7 off the trailer. He began to assemble the ship as he had planned to fly it first that afternoon.

I encouraged him every way that I could by agreeing to let him fly it first. On the completion of one of the tows, I saw Wally come rushing over towards me, wildly waving his hands at me.

Strange, I thought, Why is Wally acting like this? I cut the engine on the Waco, and Wally very excited said, "Mark, I think you had better take a minute to come over and look at your BG-7."

"Why, what is the matter?" I replied.

"You had better look for yourself," he replied.

"OK, if you insist Wally, I'll go over there now."

I was completely unprepared for what transpired next.

"Wally, what the hell did you do to my ship?" I moaned.

"Nothing, Mark, that's the way I found it, I didn't do a thing until you saw it."

The top hinge, along with the middle hinge, had broken completely off the post. "I don't believe this, Wally. I just got this damn rudder repaired and the ship relicensed."

"I know," said Wally, "but it looks like the guy brazed the fittings instead of welding them. Well, that does it, no way will we be able to fly this ship during this meet."

I was mad as a wet hornet, and stormed up to the main hangar to see if I could possibly get an FAA mechanic's opinion. I knew "Mac" the local FAA mechanic, and asked him if he would check it out.

"Be glad to, Mark," said Mac. "Bring it on up here to the hangar."

Hooking the trailer and glider up again on the car, we brought the sick bird up to Mac.

"Well, it's bad, but not hopeless," said Mac.

"Do you think I might still fly it for the meet, Mac?" I asked.

"It will be tough, Mark, I'm not saying no, but I personally would not try to do it," replied Mac.

After talking to Mac for a while, he agreed to weld up the fittings if, I agreed to get it ready for him.

"Sure," I said. Now with renewed hope, I cut back the fabric to expose the damaged area. With Wally's help, we stripped and sanded down the tubing prior to welding.

"Remember, Mark," said Mac, "all the surfaces must be absolutely clean in order for me to get a good weld."

Wally and I worked for over two hours getting the rudder post and fittings ready for the torch. We did take an occasional break to go out and watch our friends soaring above us with their ships.

Mac finally said, "OK, it looks fine, you had better get plenty of wet rags and water to keep your ship from burning up, as I weld."

Our next step was to super-soak the area by keeping wet rags over the fabric.

With several fire extinguishers standing by, Mac started up his gas welder and began to repair the damaged parts. It took about a half hour to weld. Finally, it was repaired, the right way.

"Well, Mark," said Mac, "your work has just begun." That turned out to be the understatement of the year.

We towed the glider back to our tent and started to repair the bird.

The first step was to prepare the surface. We thoroughly cleaned the weld and surrounding areas and put zinc chromate on all the bare metal parts. Time was slipping away. We had skipped lunch, and now supper. By now, another friend, Ralph Turner had stopped by to see how we were doing. Naturally, we pressed him into service working with us, Wally and I started to sew up the fabric. Using the standard baseball stitch, we soon had the job completed.

Incredible! We could hardly believe our eyes. The time had slipped by to where it was almost 9 PM. Everyone by now had heard about our plight. Each one in turn made a particular point of coming over to our tent to offer their condolences and pay their respects. We believed it was more out of curiosity than anything else.

It was close to midnight as we started doping the fabric with the first coat of dope. "One down, three to go," said Wally.

"Not really, we still have to get the final silver coat on after the doping," I replied.

"Damn," Ralph said, "it's starting to rain."

"Oh, no! This will really put us in one hell-of-a-bind," I moaned.

Wally said, "Hey! I got an idea. Let's put the tail in our tent, and use our Coleman stove and lanterns to help dry it quicker."

"Good idea," was our joint reply, hoping that we would not blow ourselves up in the process.

This helped tremendously, as the first coat dried in about an hour. More visitors, more ideas and a lot of smart ass remarks such as: "Boy, you guys have the biggest tail of all in any tent around here tonight." Ignoring their tauntings, we labored on.

Many of our visitors left the tent in less than five minutes, eyes full of tears, and gasping for fresh air.

Now it was close to 2 AM and we still had yet to put on the required extra silver paint coat. We had planned to sleep six guys in this tent, but everyone was driven out by the smell and fumes. That is, everyone except one guy named George Bailey.

About 3 AM, he finally surrendered and left with his sleeping bag and pillow under his arms. He elected to sleep under the trailer parked outside of our tent.

We heard the rain beating a steady tattoo on the tent for the past hour. It slowed our efforts, but we were determined to finish the job.

It was six o'clock in the morning as we three (Ralph, Wally and myself) staggered out of our tent. Bleary-eyed, wheezing, and gulping for fresh air. We had beaten the odds, and made liars out of a lot of pilot friends who claimed that there was absolutely no way we could get this ship ready for the contest.

Incredible! There was George sound asleep under the trailer. It was pouring down rain and puddles of water had formed small little lakes all over his sleeping bag. George was oblivious of everything and everybody.

"Hey George! You're all wet!" we yelled. George woke up with a start and discovered he was wet clear through. He thought that the trailer would offer some protection, but in the dark did not realize that it was an open trailer with no deck cover.

"Well, George, you missed all the excitement," I said, "the patient will live to fly again."

Later that morning, the rain finally let up and the weather started clearing. Unfortunately it was too late for a contest day.

This gave everyone the time to come over to our tent and pay their respects to our dead bird.

As the group swelled from a couple to over twenty, it soon became obvious that the somber mood had dramatically changed.

Everyone had been convinced that that was absolutely no way that we could possibly compete in this meet. We even had some offers from many of our pilot friends to let us crew for them.

Of course this would mean no flying and all work as we went out over the boon-docks trying to find them. Then we had to disassemble their bird, load it on the trailer, drive back to the Richmond Airport, and reassemble the ship. We enjoyed the competition and were willing to help. But this would be ridiculous. We came here to fly in the competition, not to work for other pilots.

"Are you guys kidding?" we would respond, knowing full well that down deep, they did have sympathy for us. It was their special way to help us keep up our morale.

The weather finally cleared late that evening and we soon had a host of friends in our tent celebrating the almost incredible fete we had just accomplished.

We soon got tired of all the good natured smart remarks made by our friends. I must admit, the one that stuck in our minds the most was made by a very good friend of mine late that following evening.

"You know Savage, I doubt if any one can top the act you guys put on last night."

"Yeah, I know what you mean," I said.

"I don't think you really do," was his response.

"Oh! What are you driving at?" I replied.

"There is no question about what had happened to you guys. All the partying and drinking that was going on in your tent last night was something else. And . . . incidentally, no one can ever say you guys didn't have the biggest tail in your tent last night."

Finally, after all the laughter died down, all I could say in response was, "You're absolutely right, we did have a wild time in our tent last night. But . . . she's up and ready, and raring to fly again."

Chapter 7
A Birthday Surprise

It was in the late fall and I had been soaring with my sailplane known as a Breglieb BG-7, over the Chillicothe hills in Ohio, when unfortunately, I had lost the thermal and had to land. I felt since the thermals were now starting to get very active, I would take another tow back to the same area I had just left.

Before I had a chance to get hooked up for my second flight, Ted Wedenmeyer, who had been flying the Waco tow ship, signaled me to come over to the tow ship. The engine was running and sounded very smooth. Puzzled, I ran over and yelled over the roar of the engine, "What's the problem, Ted?"

"Actually, nothing," he yelled back. "She's running like a fine tuned watch. So, how about taking a few tows for me, while I grab a coffee and a sandwich?"

"OK, I'll be happy to take a few and after you get your break, you can tow me back up," I replied.

Our usual practice was to leave the engine running. He climbed out of the cockpit and I climbed in. I visually checked the gas sight gauges, and everything looked fine.

"Two and a half hours of flying . . . the sight gags showed about one quarter of the fuel down . . . that's about normal," I said aloud to myself, as I finished the check list.

The final resting spot of the Waco UPF-7 as it appeared in the local Chillicothe paper.

I shouted at Ted as he stood by my wing, "How's the engine running?"

"Fine," he shouted back to me, as he made his way over to the local hamburger stand.

So far, we had put a total of twenty hours flight time on this recently overhauled engine. I had helped a couple of other guys run a top overhaul on it, and I was concerned that nothing unusual was developing.

The air had good lift and the Waco Bi-plane towed like it had no load at all. The next tow went better than the first and it felt good to know that the recently overhauled engine made this much difference.

The third tow was with a Schweizer 1-26 single seat sailplane with Harry McIntyre as the pilot. I had taught him to fly during the past year and he always tried to get as much as he could from every flight he made in the glider. Can't say as I blame him.

He had sent word up to me that he wanted a direct up-wind tow. In all of our past operations we always agreed to stay away from making any direct tows over the city. If I honored his request, I would be flying directly over the city of Chillicothe. This was completely against our general rules. But, today the wind was blowing directly out of the southwest about 20 knots. This put our flight paths directly over the city if we towed into the wind.

I felt that as good as the Waco was towing, I should be at least 1500-1800 feet over the southwest end of town. No question, this would be a safe altitude for the glider just in case of a rope break.

With this kind of altitude, Mac could easily stand a better chance for a longer flight. He would be far upwind to start with.

I thought, if I towed him in the normal pattern, his drift from the strong winds would take him away from the field and shorten his flight considerably.

With all that rationalization behind me, I got the signal from the ground crew and added the power to start the tow. I checked the pattern and started my take-off. WOW! The old gal was really doing its job. We actually leaped into the air in one of the shortest runs I have ever had.

I visually checked the sight gauges once again. Strange . . . they hardly moved at all. We had one sight gauge located on the bottom section of each of the upper wings. The gauges were quite similar to the old Ford Model A's of years gone by.

Well, with only a dozen tows so far, and five hours of fuel in the tanks, I doubt if the gauges would give me much of a big indication.

"Hmmm, according to my calculations, the sight gauges are just about right where they should be for the short amount of time we have flown today," I say to myself.

Then at 1200 feet . . . POW! Suddenly it is extremely quiet. My first reaction is to release the glider. I reach over and pull my lever to let the rope drop from me. Next, I half crawled inside the fuselage cockpit and reached down to change the gas selector valve. The engine sputtered back to life with a bang . . . but only for a few seconds.

I switched to the other fuel tank. The engine came back to life. But it died once again just as quickly as it had come back. Only this time, it did not start back up. No doubt about it, I had fuel starvation.

That's strange? I still was showing I had at least a half full indication on both sight gauges. "I can't believe this," I moaned. I cursed briefly saying, "Oh, shit! I just can't believe this is really happening to me."

I looked around and could not find a suitable spot to land. "Boy, I'm glad it's me here, rather than some other member who had low time, and experience," I said aloud to myself.

I looked around again, only this time more carefully. I still could not find a single place to land.

Oh Lord, there is a golf course off to my left. I wonder if I could glide this thing that far?

"I'm afraid not, Savage," I told myself aloud. I still vividly recall the day that young kid was killed on the Salerno Beachhead airstrip in Italy, when he tried to stretch his glide.

Trees, houses, poles, wires, were everywhere. Everything was cluttered up. Damn . . . why did I break our rules about towing over town? Not a field of any kind in sight.

A cartoon that appeared in "Flyer Magazine" that depicted my exact feelings after the crash landing.

"Why, oh why, did I ever get talked into towing Mac over town?" I say again to myself. "Well, it's too late now, and I had better keep my cool and settle down," I kept saying over and over to myself, while frantically looking for a place to land.

There, over to the left, is a golf course. Damn! that's the same one you saw before, Savage, and it's just too far to glide there safely. Maybe if I stretched my glide I might make it, but, I could also spin in if I get too slow, and then try to bank in for the landing.

Frantically I looked everything over again for the umpteenth time.

The only possible and logical landing area was an orchard located on the side of a hill!

That's it! My decision is made, I have no other choice. I will plan to land in this orchard and hope that I won't get impaled as I plant it on top of the trees.

I lowered the nose of the Waco and made a very shallow turn towards the orchard. I shut off all switches. I reached down and tightened up my seat belt and shoulder harness. I tried to make it as tight as I could.

Meanwhile, Mac off tow, dropped the tow rope and circled over head and watched the terrifying, dramatic scene unfolding before his eyes.

The wires on the Waco's wings were screaming like some wounded banshee. I was dropping like a rock. Suddenly, as I flew over the high-tension wires, trees, and houses I spotted a very small field ahead. Damn! This field couldn't have been more than six hundred feet long.

I put the Waco Bi-Plane in a violent full left slip. Probably more from instinct than anything else. I kept thinking. This just can't be real, it just isn't happening to me again.

I have had several exciting experiences in the Air Force in combat with two forced landings in one week. (Believe me, this one was no different). I must try to keep my cool and stay calm. Above all . . . I don't want to panic! The worst thing that could happen to me now would be for me to quit thinking and give up.

I came screaming over the high wires and trees and cleared

them by a mere breath, as the ground came rushing up to meet me.

I looked and remarked aloud, "Boy, that landing area is so damn small, I'm not sure I can make it."

I put the Waco in a big side slip, and was not worrying about making a good landing. I could only hope and pray that it would not bounce. If I bounced, I could skip up and start flying again right through this garage that loomed up in front of me.

The left wing and nose smacked the ground with a resounding thud! Crunch! Went the left gear and prop. The ship spun around to the left. Fortunately, I didn't bounce, but she sure hit the ground mighty damn hard.

The force of the impact sent me up against the seat and shoulder harness. The straps were fastened over the back of the metal pilot's seat. The impact force abruptly pulled me forward.

The back side of the metal seat bent almost completely in half from the tremendous force of the weight of my body and the shoulder harness over it.

My head snapped forward and I made contact with the cockpit leather sponge combing that I had recently installed.

Man, oh man! I was down and safe! I think.

As I shook the cobwebs out of my head, I looked around and felt sick. Our CSA (Columbus Soaring Association) beautiful bird lay on its broken gear. The lower left wing was shattered. The left main landing gear had been knocked off and lay under the crumpled wing. The prop was bent under the engine, and the right wing tip came to rest against this guy's garage.

I stare in amazement with my right eye, as a huge bump started to swell up over the left side of my face.

I said to myself, "Oh God! what a hell of a headache I'm going to have. But at least I'm alive!"

Later that day, I found out that Mac had circled over the crash site and then flown back to the airport to report the accident to the club members.

As I climbed out of the cockpit, I was astounded to see a big white ambulance waiting for me.

The repaired Waco UPF-7 flying nine months later after the serious crash in Chillicothe, Ohio.

"Odd . . . How did I get such fast service? Was I knocked out that long?" These and many more questions kept popping up in my mind.

The attendant asked me, "Sir, would you like to lay down on the bed?"

"Hell no!" I replied. "If I'm going to die, it's going to be sitting up."

With the red lights flashing, siren blaring, we roared off to the hospital. Strange? We had not seen a single soul as we raced off for the hospital.

At the hospital, a nurse wrapped a blood pressure cuff around my arm and then after pumping it up asked, "Looks like you will get quite a shiner from the bump on your face below your left eye."

"Just how in the world did that ambulance get to me so fast?"

"Oh," she said, "The reason for the fast ambulance service should be credited to an ex-Air Force flight nurse."

"I don't understand?" I said.

"Well, it seems that she always watched your ships flying over the Chillicothe area from her home located on the crest of the hill where you crashed."

She continued, "only this time she heard you falter and knew that it was inevitable that you were about to crash. She called the local emergency squad, and they were there on the spot to pick you up just about the time you crashed."

All I could say was "Incredible!"

"Now," she exclaimed, "I just can't believe it. You are so calm and checking out at 140/80. Why that's perfectly normal. Did you really crash an airplane?"

After completing all the X-rays, it looked like the only serious damage to me is a tremendous bump under my eye, and a busted pride. Within the next few weeks a colossal black eye replaced the huge bump.

I arrived back at the crash site several hours later and I'm greeted by just about the whole town of Chillicothe. They are trampling all over this guy's grass and bushes trying to get a glimpse of the wrecked airplane.

The man who owned the house where I crashed was actually having a surprise birthday party today. Unfortunately, they hadn't planned on this airplane and me as their big surprise.

Chapter 8
Three Into Twelve Equals One

In order for anyone to solve this impossible equation, it will be necessary for me to turn back the clock several years. It all started when a good friend of mine named Wally Marsh, began flying my Breigleib, BG-7. He liked the performance so much, he decided that a new sailplane kit called the BG-12 would be that much better and just the ticket. He planned to build this sailplane in order to make long cross-country flights and compete at the various glider meets. In addition to the meets, he wanted to work towards his F.A.I. badge awards. These badge requirements were set up by the French many years ago to serve as a standard throughout the world. It deals with altitudes, speeds, and distance flights.

Wally and Ralph Turner, were a couple of former students from OSU who had graduated and were now working together as Aeronautical Engineers at the North American Aviation plant in Columbus, Ohio. As a result of the former ties with the gliders while attending OSU, they decided to form a partnership and buy this BG-12 sailplane kit.

Ralph would build the fuselage and tail, while Wally would tackle the wings and center section.

It took them about two years to finish this project. Wally

The Breiglieb BG-12 with its three builders. Jim McLurg on the left, Bob Kosar seated and Mark Savage on the right. A Schweizer two-seat 2-33 trainer sits in the background.

planned to build the wings in his warm and comfortable basement.

When it came time to get the wings out, he had to remove part of his basement wall in the process. This strange building and removal procedure had captured the interest of a local reporter and his unique story appeared as a feature article in the Sunday's paper.

Several years had passed since this ship was built. Wally and Ralph had accumulated enough flying time that they both felt quite comfortable flying their jointly owned and built BG-12.

One particular week-end in the fall the weather forecast promised excellent soaring conditions for flying sailplanes. Wally had contacted me earlier that week and asked me if it would be possible to meet him at the Marion Municipal Airport someday to give him a tow providing these conditions held true for that week.

"Of course, Mark, I would like to get an early start. I hope to be able to fly far enough to get my gold distance badge. I'm planning to fly to Lexington, Kentucky," said Wally.

"That's great, Wally," I replied. "What would be a good time for me to meet you at the airport?"

"I would like to get airborn by ten if you think that's possible," said Wally.

"Sure thing," I replied. "See you then."

I drove Wally's car as we towed out his BG-12 to the flight line. He had a multitude of jobs to complete before I could tow him off for his flight. I agreed I would crew for him, and stand by the phone for his call.

I planned to drive his car and glider trailer to retrieve him from wherever he should happen to land.

I towed him up to 2,000 feet above the ground shortly after ten and watched him depart on this, his proposed long cross country flight.

It was shortly after five that day when I received a call from Wally requesting me to come down and pick him up.

"I'll be happy to Wally, but, just where in hell are you?" I asked.

Then Wally, by his own admission, said, "Mark, complacency had set in with disastrous results."

Fearing what he was about to say, I asked, "What do you mean by that statement, Wally?"

"I'm down here at Rising Sun Airport in Kentucky. Both the BG-12 and I parted company," he replied.

I tried to get him to tell me more, but he avoided my questions and asked me to please start down to Rising Sun as soon as possible, as it would be very late by the time we would get back home.

I felt we would need extra help, so I drafted a mutual friend of ours, Jim McLurg, to drive and help. It was quite a job for three people just to get the BG-12 sailplane de-rigged and on the trailer.

Jim and I packed up Wally's car and trailer. Jean, Wally's wife joined us as we departed for the Rising Sun Airport to pick up Wally and his Twelve. We arrived at Rising Sun just about the time the sun was setting.

"Ironic, isn't it Jim? Here we are arriving at the Rising Sun Airport just about the same time the sun is setting. Do you think they should have called this the Setting Sun Airport?"

"Yeah, if you say so, Mark," replied Jim. We were both tired and my remark did not go over very well.

We met Wally at the entrance to the field and he was carrying his chute. We all noted he was not his usual self.

He was very somber and his clothes were covered with dirt. He also had a small scratch over his one eye.

"Good God! Wally. What in-the-hell happened to you?" I asked.

He replied, "Despite the fact I had about 5,000 feet of altitude, and four hours of daylight left, I mis-read my maps and thought it was too late to make my original planned long distance flight.

"Well, Wally," said Jim, "Mark, Jean and I, know it's a hell of a lot further by car. It took us almost four and a half hours just to drive down here."

Wally ignored Jim's remark as he continued, "Actually, I was just a mere 25 miles away and could have easily made a long let down glide to my destination. Instead, what did I do? I circled over the Rising Sun Airport for almost two hours before I elected to land."

"You look awful Wally, so what happened to the 12?" I asked, fearing the worst.

Wally ignored my question as well, continued. "After many circuits of the field checking out the area completely, and noting the wind and wind sock direction, I elected to land on the taxi-way of this grass strip airport. I had hoped to keep the work for my crew down to a minimum."

"OK, we understand Wally. But, what about the twelve?" I asked again, impatiently.

"On touch down, the right wing tip splate caught in high grass."

"So what happened next?" asked Jim.

"It was just like a slow motion movie show. All I could do was sit and ride it out. As the ship came to a jarring halt, I was ejected out of the cockpit like a pea out of a pod. The fuselage and tail section were completely demolished."

Jean was on the verge of tears as Wally continued his story.

"The wings survived along with the bubble canopy. My only injury was to my pride."

"Oh, no! I just can't be," said Jim and I together. Then Jean burst out in tears.

As a result of this late fall accident, the brand new partnership of Marsh, McLurg and Savage was formed and we bought a new fuselage kit.

We all three felt that if we worked together the Twelve would be flying early next year, or fall at the latest. How wrong we were.

Wally had moved into his new home that happened to have a two car attached garage. This, we decided, would give us an ideal hangar and place to work.

A brand new DeWalt power swing saw was the most valuable tool we had. Wally was a real wizard on this machine. As a result of our combined efforts, the newly purchased kit with the fuselage and tail components began to take shape rapidly.

Early one morning, a woman was driving past Wally's house. She was late for work, and for some unknown reason lost control of her car. Yes, you guessed it, she left the road, jumped the ditch, and ran over Wally's new lawn and hit the garage dead center. What a crash!

The finished wings were pushed about two feet off their stands. We were very lucky though, as they only received minor damage. The biggest problem was the garage. Our work on the Twelve came to a screeching halt. We had to wait a couple of months, until the garage was repaired.

Despite this delay, we did achieve some real benefits from this catastrophe.

Wally received enough money to help get the garage back in shape in a relatively short time. He had time to do a bit of head scratching. He figured if the fuselage had been double planked this ground loop would not have spelled disaster.

This new design work proved agonizingly slow due to all the many cuts and fits that had to be made on every section of the fuselage. Several months later, the job was completed.

It turned out beautiful and strong. Including all this labor, much time was spent in the discussion of trying to improve the performance of the Twelve.

We all felt that the Twelve had a decided problem with directional stability. The biggest problem of all was excessive pitch.

As a result, an accidental side benefit evolved from this change. We could never trailer the Twelve with its tail feathers on. Originally the ship had exceeded the maximum Ohio road legal width limits. Now, with the new designed tail, compliments of our Aeronautical Engineer, we are able to leave the tail and elevator completely assembled for trailering. Another plus benefit was the amount of time spent on rig and de-rigging. With this new fix, the total time had been cut almost in half.

Our bird was taking shape. It was indeed quite different than any other BG-12 in the United States.

That following winter, it was decided Jim and I would work on the new fuselage and install flaps, seat, and flight controls. The surviving canopy would be the last item mated to the new fuselage. We both decided the nose should be redesigned with a new fibre-glass nose cone.

While Jim and I worked diligently on this project, Wally agreed to finish the wings.

The next question was where to work? Jim lived out in the country, but had no real place to work. An old friend and former glider pilot, Harry McIntyre, just so happened to be building a magnificent ranch home near Jim. Although no longer active in the soaring club, he agreed to let us use his living room to work on the Twelve.

"Kind of neat, Jim," I remarked.

"Imagine, building a sailplane in a house that was being built."

Yes, the neighbors did shake their heads on this one. In fact, it was so unique, one of the local TV stations had heard about us, and did a five minute program on our joint venture.

One night, during the refreshment break, I heeded the call of Mother Nature. Unknowingly, I had forgotten the fact that the front porch was not finished. It had been roughed in with block, but had no floor. The dirt floor was about 15 to 20 feet below the front door.

I told Jim, "I'll be back in a couple of minutes," as I opened the front door and took one big step . . . down!

It was pitch dark and I thought I would never stop falling. Luckily, I held on to the door knob just long enough to help slow my fall, as the door slammed shut behind me and I crashed on the dirt below. It was pitch-black and I was in a state of shock. I was numb and I tried to feel my body to see what I might have broken.

A few minutes had elapsed when Jim began to wonder what had happened to me. In shock, Jim opened the door and stuck his head out and yelled down to me. "Savage, now just how the hell were you able to shut that door?"

"It wasn't easy," I replied.

"Oh yeah? Is that right?" he replied, as he closed the door and went back to work. The whole area turned pitch black once more.

Countless evenings of cutting and welding and the front end began to take shape. We had finished the fuselage and tail and it looked great. It was time now to get started on refinishing the wings. We jointly decided to lightly sand the wings and center section, and then paint them when we painted the fuselage.

After a few nights of sanding, it soon became evident that this would not be the best thing to do. What began as a refurbishing job, wound up as a major project. What with all the effort expended on the Twelve already, it was a unanimous decision, we just had to redo the wings and the center section. It was felt that, with a little effort, we could get the desired finish on the wing sections.

The more we sanded, the more convinced we became that the old fabric finish on the wood was not good enough. A week passed and the moment of truth arrived. The old cover had to come off.

We stripped, sanded. Stripped, sanded, add infinitum. It seemed to take forever. Before long, we had many rivers of white filler liquid flowing down Wally's drive and out into the street as we washed down the wings. It was a tough task pleasing three guys, each one wanting the job to be perfectly ultra super smooth. Eventually the wing area filling was completed.

With all the new advances in glass technology, we knew that the right decision had been made to glass our wings. Finally, the ship was ready to paint. We had no painting facilities, so it was decided to farm this out.

This was one of the best decisions we made. The final inspection was approved by the FAA (Federal Aviation Administration) inspector.

The date for the big day arrived two years and six months beyond our original estimate. The Twelve with all the changes turned out to be a magnificent flying machine. The ship has an almost hands off flight characteristic. Now with the spanking look of a glass ship, it is preened and ready to wear the diamonds it so rightly deserves. (Diamonds are the awards given to pilots who have achieved set tasks as required by the FAI (Federal Aeronautic Internationale) So, by now you should have the answer to the problem. Three into a BG-12, equals one beautiful bird.

Chapter 9
"May Day . . .
May Day . . .Mid-Air!"

As these terse words crackled out over the air, nearly everyone who heard this call literally froze on the spot. That particular day dawned fair with the promise of excellent soaring conditions. The Central Ohio Soaring Association has operated at the Marion Municipal Airport at Marion, Ohio for over 10 years. It has enjoyed a good home with excellent club growth. COSA, as it is more popularly known, has more than 90 members on its roster. The club equipment consists of: 2-two seaters, and 2 single seat ships. One particular single seat sail plane was known as a Schweizer 1-26.

All of the sailplanes are launched by a standard Cessna 150 two seat trainer that has been converted to a 150 horsepower Lycoming engine. In addition, a special climb propellor and droop wing tips have been installed for better performance.

This tow ship can get just about any sailplane launched up to 2,000 ft., and be back down on the ground in about 8 minutes. Over the many years of operation, the club has made more than 20,000 tows with the club's tow ships.

Despite the fact the ground is relatively flat, this area produces a lot of good thermals in the summer. As a result, many pilots have made some outstanding flights.

A Schweizer 1-26 sailplane similar to the one flown by Linda Huelsman. Her husband's sailplane, a BG-12, collided with her sailplane at 6,000 above the ground and ripped off her left wing causing her to spin all the way to the ground.

There have been several diamond badge flights recorded from this airport. A complete Diamond Badge consists basically of three awards, but first a pilot must have completed one flight of at least five hours duration.

One leg is an altitude gain of 5,000 meters or 16,400 ft. The second leg is a goal flight of an out and return triangle of 300 km., or 186.4 miles. The third leg is a distance flight of 500 km., or 310.60 miles. As one can see, this is quite a feat.

Two particular club members, Linda Huelsman, a piano teacher, and an avid soaring nut, along with her husband Neil, an insurance salesman, were to have an experience that will live in the member's minds forever.

Cumulus clouds were starting to form around 11 AM that day, and Linda requested the Chief of Operations to schedule her a tow in the club's Schweizer 1-26. This was a single seat sailplane. She had high hopes of being able to get her 5 hours duration today to complete the requirements of the FAI, Silver "C" award.

Linda was still a student pilot but quite competent, and very close to getting her Private Pilot Glider license.

Linda carried a club CB in her sailplane and prior to her launch, it was agreed that if any member wanted the club ship, they could call on the CB radio and advise Linda to come in and land.

The launch was uneventful and soon she was soaring around 5,000 feet above the ground.

With such good conditions, several other members were soon airborne in their sailplanes enjoying the pleasures of a day that promised excellent soaring.

Meanwhile, Neil had been busy doing some minor work on his ship, a Breiglieb BG-12. Linda by this time had over 2 hours and had radioed down to Neil advising him. "The conditions were just great."

Neil, alerted by his wife's remarks, hustled to get ready and was launched in less than five minutes. He released at the customary 2,000 feet, the club's regular altitude release point. He immediately caught a strong thermal and was soon able to join Linda under her cloud in this large thermal.

She called to him on the radio again, advising him that she was about 200 feet above and behind him.

Neil acknowledged her call as he made his turn and entered the thermal in the same direction that Linda was turning.

Meanwhile, on the ground more members were showing up and wanting to fly. The club rules for single seat ships was a maximum 45 minute time limit for Saturday and Sunday.

After a brief meeting, it was agreed by the Operations Officer that since Linda was flying a club ship, she should come down.

Russell Savage the Operations Officer for the day, picked up the mike to make the call on the base CB advising her that the waiting list had grown and she should return to the airport as soon as possible.

Flipping on the radio switch, the very first words that came over the air was, "MAY DAY! . . . MAY DAY! . . . MID-AIR COLLISION AT 5,200 FEET! . . . MAY DAY! . . ."

Everyone on the ground was stunned by this transmission. Again came Linda's pleading voice, "I AM UPSIDE DOWN! OH MY GOD . . .! THE LEFT WING JUST CAME OFF . . .!"

A dead silence . . . then . . . "Someone . . . please help me . . . now I am spinning towards the ground."

Another totally unknown voice cut in and said, "Everyone stay off the air, this is an emergency."

Everyone, including Russell, were totally shocked by this last transmission.

Russell, keyed his mike button, gulped, and said, "Where are you? What is your position Linda?" hoping that she might hear and answer his call.

Finally after what seemed like an eternity, Linda's voice boomed out over the radio. "It looks like I might crash in a bean field about three and a half miles from the airport."

One of the club's 2 place Schweizer 2-33's was flying in the vicinity of the airport with instructor pilot, John Luepp and his student, Jeff Edgell.

Jeff remarked to John, "I thought no club ship was allowed to do acrobatics?"

"They're not," replied John. "Why do you ask?"

"Look! over there. It's a 1-26 sailplane spinning," replied Jeff.

As they watched this maneuver, John wondered when the pilot would come out of this spin.

Their complete attention was focused on this ship as they watched in horror. The 1-26 was continuing its spin down towards the ground.

"Good God, it just hit the ground!" Jeff shouted.

"I know, I know," cried out John, as he grabbed the controls from Jeff, his student. Applying full spoilers, he dove down to the airport in order to try and summon help.

Neil in his BG-12 circling to the left first saw Linda loom in on his left side. He had gotten the stronger edge of the thermal and had gained the 200 feet difference in less than a half turn. Horrified, the first thing he saw was his left wing contact the left wing of his wife's 1-26, then it tore through her wing. His left wing had ripped off her left horizontal stabilizer and elevator.

The next thing he knew was the instrument panel had popped off on to his lap. He was now in a screaming 140 MPH dive. Gingerly he eased back on the stick and found that he still had some control left. He was just barely able to fly his BG-12 back to the field.

Landing, he shakily climbed out of his ship and was absolutely speechless when he found over 8 feet of his left wing and part of his aileron missing.

A group of members rushed over to him and were amazed to think he had actually made a safe landing. About all Neil could say, with eyes full of tears, "I killed my wife! Oh God! I killed my wife!", over and over again.

Linda was now in a ship that was completely out of control. The Schweizer 1-26 was in a flat spin to the left.

She did remember seeing the sight of the top of the trees. This indeed was an indication that the end was near.

She tucked her left foot up towards her body, then pushed with all her might on the right rudder pedal.

She held the Stick full back, and pulled open the spoilers.

Now she believed, and was positive, that everything she

had been taught would help save her life.

Slowly . . . ever so slowly . . ., the ship stopped spinning. It was almost at that exact second the ship hit the ground with a sickening thud.

The soy bean field had acted as a cushion and despite all the crunching of metal, the ship's cockpit stayed intact.

Linda could hardly believe it. She shouted with joy as the farmer-owner came rushing over fearing the worst. She had a few bruises on her knees, and a small cut on her head.

Now the emergency squad, summoned by the alert and unknown CB voice, arrived on the scene several minutes later.

Meanwhile, Neil had another club member, Jim McLurg, get his car in order to help him drive over to the crash scene. At this point in time he felt that he was in no shape to be driving. Both he and Jim knew full well what they could expect to see when they got to the crash scene.

Jim put the red Capri in low gear and let it rip. The tac rose to a screaming 6500 rpm, well above the red line. This was the highest Neil had ever seen it go. It was a wonder it didn't blow up the engine. Wheels spinning, the Capri flew off to the crash site in a few minutes.

Neil, weak-kneed, sobbing, eyes flooded with tears could only keep saying, "I killed my wife," as the car stopped in a huge cloud of dust.

As the dust slowly settled, Neil and Jim were shocked.

There was Linda sitting in the ambulance.

Jim shouted, "Neil! . . . she's alive!"

A joyful reunion, but now it was time to go to the hospital. She did not want to go, but complied with all the pleas that she surely must have sustained some type of injuries and she should be examined by the doctor.

At the hospital, Linda joked and explained how she owed her life to the excellent training and the sturdy Schweizer 1-26 that helped keep her from leaving this world.

X-rays, blood pressure, etc., all checked out just fine.

The proper authorities were advised and soon the NTSB (National Transportation Safety Board) rep and the local FAA personnel arrived and started piecing together all of the facts.

The BG-12, a higher performance ship had rapidly outclimbed the 1-26. Despite the communication acknowledgments, the extra lift had caught Neil by surprise. With both ships in a standard left turn, it was inevitable that unless a wider turn was made, a collision would occur.

Linda saw Neil's wing looming larger and larger. She stared at it with hypnotic eyes. But, she could not out-turn the faster ship. All she could do was wait as Neil's left wing cut across her left tail and wing.

This was incredible. The impact of the collision caused her left wing to actually bend over double and the bottom of her wing struck the main tire and left a big black mark on the bottom of the left wing panel. Linda's ship immediately went into a dive.

At about 5,000 feet, the left wing separated from the ship and floated away. The 1-26 actually maple-leafed all the way down to the ground.

Later investigation showed the major part of the left wing had landed about a quarter mile from where she crashed.

In retrospect, this was an accident that will go down in the history books as being incredible and impossible for anyone to survive.

There is no doubt that the sturdy American-built Schweizer 1-26 helped her survive. Linda later came back and offered her praise and congratulations to me for teaching her how to fly and recover from spins.

"You know," she said, "I remembered everything you taught me about spins and spin recovery, and that's what helped me the most. I didn't panic, as I heard you talking to me, saying do this and do that," replied Linda.

All I could say at this point in time was, "Linda . . . maybe . . but really . . . I'm sure the good Lord just didn't want you . . . yet!"

Chapter 10
"This is an Emergency"

After graduation from K.S.U. (Kent State University), I was offered the job of working for the State of Ohio with the Ohio Aviation Board.

I knew it did not pay much money, but the opportunity to fly again helped me decide rather quickly. The State had two Cessna 170's. It was not a fast ship, but well suited for our purposes. I was placed in charge of the airport inspection program. In conjunction with the inspection program, I did some flying for other departments in the State.

Over this past year, we had been working rather closely with the Department of Economic Development. One of the best slogans that I ever heard, came from there. It was, "Ohio . . gave Flight and Light to the world."

Indeed it did, with the Wright brothers, Dayton, Ohio and Thomas Edison, Milan, Ohio.

We flew all over the state and sometimes took some of the key people from the Economic department with us. One guy a good friend of mine named Al Green, was scheduled to fly with me from Columbus, to Painesville then on to Toledo. Actually, he only wanted to go to Toledo and return.

Our department tried to double up on trips in order to save

This venerable Cessna 170 formed the backbone of the Ohio Aviation Board activities. It was used for the on-going airport inspection programs and host of other Ohio aviation related activities.

money. After he heard my simple explanation he agreed to fly up to Painesville with me and then on to Toledo.

Why? Because I had convinced him that flying was faster, safer, and a lot more fun.

I assured him that I would wait for him until after his meeting was over in Toledo. Then we would have a nice leisurely flight back to Columbus. Actually he was only interested in getting from Columbus to Toledo and return. Now, with an extra leg thrown in, this became a rather long flight and would probably take most of the day to complete.

Al had never flown before and I went out of my way to assure him there was absolutely nothing to worry about. It was early spring and the air was rapidly changing from winter to spring. A beautiful spring dawn beckoned us into the air for this flight. At this early hour of 7:30 in the morning the air was just as smooth as glass.

Since this was Al's first flight, I had to remind him that the air was not always this smooth and warned him by saying, "Probably this afternoon it could be a lot rougher."

Al came back with, "I think that this flying is OK, but don't get me wrong, I would rather be driving."

"You can't really mean that Al," I said, "look how fast we're going, no traffic cops to worry about, and no rough roads."

This particular leg was very fast with a strong tail wind pushing us. Painesville is located in the far northeastern corner of the state.

Al constantly reminded me that all he really needed to go to was Toledo, not Painesville. This was exactly one hundred and eighty degrees opposite his destination and almost as far away as Columbus was to Toledo. Toledo is located in the western edge and opposite corner of the State. He couldn't understand why I had to go to Painesville first with Toledo our next stop.

"But isn't that way out of our way?" he asked.

"Well by air, not really," I assured him, as we landed at Painesville in record time, just a shade over thirty minutes. I hated to tell him, but our next leg would be at least two hours, as we would be fighting almost a direct head-wind.

I had the crew refuel me while I was at my meeting that took less than twenty minutes. We were airborne in a little over a half hour. Just about as long as it took us to fly up there. The air had gotten rougher and I advised him: "I'll fly directly towards Toledo. This will save us roughly about a half hour or so of flying."

"That's great," he replied, "but won't that put us out over Lake Erie?"

"Why, yes it will, Al. But, we shouldn't be more than a few miles out over the lake shore at the most," I replied. "In fact if any thing should happen we will be able to easily glide to land."

"Oh no!" he cried. "What if we don't make land. I can't swim."

It was early spring and the lake ice was just breaking up and large chunks of ice floes dotted the lake and were visible in all directions.

With the wind on our nose, a bit of rough air, and out over the lake, Al was more than just a little bit uncomfortable. He was actually scared and half-sick with fear.

"Al," I said to reassure him. "Don't worry, the reason I'm flying so low and over the lake is because of the wind and the rough air."

We finally got to Toledo in a little over two hours. Both of us were relieved to get back on the ground. Each for a different reason.

Al concluded his business and we were soon air borne and on our way home. "This time," I told Al, "We will fly higher at 9,000 feet to avoid some of those extreme bumps. Of course it will be a slower trip now because of the direct cross and head-winds."

I might as well been talking to the airplane as Al hadn't heard a thing that I had said.

Finally he moaned, "Oh no! I think I would have been a lot further ahead driving this trip by myself."

"How can you say that Al?" I retorted. "Just look at all the beautiful scenery you have seen, including the lake with all the ice that was breaking up. It was a spectacular view over Lake Erie."

"If you say so, Mark. But that's not really the way I feel about this flying bit," replied Al.

"We had been flying for over three hours and are just now leaving Toledo for Columbus," said Al. "I could have almost driven it as fast as it took us to fly there."

"Nonsense," I snapped back. "Just think what a great trip this has been. Soon we will be safe and sound at the OSU Airport in Columbus, Ohio."

"Yeah, that's what you think," said Al sounding quite weary and worn out from this easy flight, we were supposed to have.

The rough air and turbulence were beginning to take its toll even on me, and I was used to flying in just about all types of weather conditions.

But, do you know? Down deep I felt that he was probably right about the time and comfort conditions.

I said sarcastically to Al, "You will never become a flyer with that kind of thinking."

The afternoon heat had gotten the air much rougher and the bumps had us really bouncing up and down. Now Al, who is black, is turning . . . pardon the expression, a little green.

I tried to reassure him that we only have about a half hour or so to go, and we will be back home. That approach didn't do much good. So I thought maybe, I had better not say too much to him at this point in time.

Suddenly, without warning, the engine quit! I reached down and pulled on the carbureator heat to check and see if the engine might have picked up some carburetor ice.

Nothing. No big deal, I thought, I'll just change tanks with the gas selector valve. Still nothing.

The propellor just kept wind milling with the engine dead. I casually tell Al, "Looks like we might have to make an emergency landing."

"Who are you trying to kid, Mark?" said Al, grinning from ear to ear.

"I'm serious Al, the engine just quit!"

"What kind of crap are you handing me now," said Al.

"Who are you trying to fool Mark. Look, can't you see? The propellor is still turning."

"Right Al, but the engine stopped, it's dead!"

"Well, that's a neat joke Mark, but you can't fool me, the propellor is still turning," said Al.

Now, I reach for the mike to broadcast my call sign to Ohio State University radio.

"OSU radio this is Cessna N-9522-A, I HAVE AN EMERGENCY! My engine just quit and I have to make a landing."

"Roger," shot back the reply. "You're cleared to use any runway of your choice, we will advise all local traffic."

"Ha, Ha, Ha, laughed Al. Mark, you'll do anything to pull off your practical jokes. Won't you?"

Ignoring Al, I said to OSU radio again, "I repeat, I have an emergency, my engine just quit."

"Roger, Roger," shot back the radio, we will clear the field and you can land on any runway."

"I would be happy too if I could," I replied. "But, I don't think I will be able to make the field from this location and altitude."

"God! You're really not bull-shitting me, we are going to make an emergency landing! We're going to die!" Al said in a very frightened voice.

Ignoring Al, I say, "Thanks, Ohio State, for your very kind offer, but I am positive I cannot make your field, even though I am at 9,000 feet," I replied.

"The wind is too strong for me to even think I would stand a chance."

"Roger, what is your last known position?" she asked in a calm, but tense voice."

Stand by, and I will let you know where I will be going down," I replied.

Luckily I found a local road map in the ship's glove compartment. Picking it up I called on the radio and told the girl exactly where I was, what road I was near, and the location of a little town of Cheshire.

I then said, "did you find that on your map?"

After a long pause, she finally replied, "I'm not sure. We will look further until we can locate your estimated landing area."

It seemed like an hour before she came back on the air with a bit of excitement in her voice.

"Roger, we found the area you told us about."

"OK, follow that road and I'll be landing just south of there on a long east/west farmer's field."

Al, by now is convinced that the end is near, and said, "Do you think we will die?"

"Not if I can help it," I assured him.

I swung the ship around into the wind for a final check on the field. Suddenly I heard another voice call my aircraft number, but it's impossible to return the call.

We are now much too low, and I am rather busy flying this ship down to an emergency landing, as I turn into the base leg in preparation for the landing.

Careful, not too slow, I keep reminding myself. That wind is very strong and it is important I don't get too slow or I could wind up in a spin.

Full flaps now. Easy, easy does it. Tail low. The field is soft for this time of the year and I would hate like hell to flip this gal over on her back.

I very gingerly pulled back on the wheel and carefully eased the craft into a three point tail wheel landing.

At this point in the game, I'm not too sure just how soft this field is, and my biggest worry would be nosing up or worst yet, flipping over on my back. We rolled less than ten feet and came to an abrupt stop. The field indeed is very, very soft.

"Whew . . .e!" said Al, "I can't hear, my ears are blocked up completely, but that's OK, we're on the ground and alive."

I shut off all the switches and breathed a big sigh of relief. Yes, it felt good to be down and safe. We got out of the airplane and faced a long muddy walk to the end of the field.

Each time I took a step, one of my boots would be sucked off my foot and left behind in this muddy field. It took us about a half hour just to walk out.

Each and every step was a tremendous physical chore, and all Al did was remind me of how stupid he was to take this little flight with me. My only thoughts at this time were centered on trying to keep my shoes from being lost in the mud.

About the time we got out of the field, Kelly, the OSU Chief Mechanic drove up to meet us.

"Boy, Mark, you gave good directions," said Kelly. "I didn't have a bit of trouble finding you."

"Say, Al," asked Kelly. "How did you like your first airplane ride?"

"First and last," replied Al, "I'll take my car from now on. I know beyond a shadow of a doubt that I will be able to make it home every time I go."

"Aw, you're still just a bit shook up," I replied. "You'll get over it."

"Yeah, replied Kelly. You couldn't have flown with a better pilot, than Mark."

"Oh? Is that right," replied Al.

"OK, you wise guys. There's another good reason for me not wanting to fly."

"Oh? There is, tell us," prompted Kelly.

"For one thing, since I can't swim, I never have to worry about flying my car out over open water."

Chapter 11
Tri-Pacer to Drummond

It seemed like only yesterday when I got a phone call from an old friend of mine, Bob Newhouse, the Mayor of Delaware, Ohio. He had been wanting to take a long cross-country trip with his Piper Tri-Pacer and was just waiting for the right time, place and opportunity.

I had been talking to him about the possibility of flying up to Drummond Island, Michigan. We had always said, perhaps someday in the future. "Well, it looks like that someday should be this week-end," Bob said, as he made this special phone call to me.

"Well Bob, we do have several guys who want to go up there before the end of this summer to see about cleaning up the place and setting up the cabin for the annual fall deer hunting season. In fact, we might even get in a little fishing while we are there."

"Yeah, Mark, it sounds like that would be a perfect reason for me to use my ship and fly up to the island," said Bob.

Well, that was all the reason Bob needed as he said, "OK, when do we get started?"

"Well, it looks like the week-end after next," I replied. "There will be four of us, counting McIntyre, Chalfont, you and

This Tri-Pacer was similar to the one owned by Bob Newhouse. The Tri-Pacer provided transportation for the four of us to Drummond Island, Michigan.

me. Four divided into the costs will make this flying trip more affordable. Doug Elliot and his dad will be driving up, so we should have transportation when we get there."

"OK," he said. I could tell he was very excited about the prospect of that long trip including the possibility of fishing in Lake Huron, and the various little lakes that were scattered around and on Drummond Island.

"Just remember, Bob, we are going up there primarily to get Mac's cabin ready for the fall deer hunting season," I said.

The weather for that particular week-end we had selected turned out to be bad. The weatherman was dishing out a lot of rain and low clouds. Since I had the instrument ticket, it was up to me to decide whether to go, or not to go.

"Mac" McIntyre and Larry Chalfont, along with Bob, were all anxious to get stated as early as we could. We had to get off early in order to give us enough daylight to land at Drummond. The airport at Drummond did not have any lights for night operations, and in addition, we all had to be back in Columbus that following Monday morning.

I checked the weather and as usual, it was bad in the Great Lakes area. One ray of hope though, the Meteorologist felt that the weather would improve the farther north we went. Drummond Island and the surrounding areas were predicted to be VFR later on that morning.

Loading up all of our gear, it looked like we were packed for at least a month's stay. Yeah, we got it all in, but it was a mightly tight squeeze. The back seat was extremcly tight to begin with.

As a result, one or the other passengers in the back seat, had to take turns putting his arm up and over the back of the seat to make enough room for the other passenger.

I filed the instrument flight plan and we were off shortly after 8 AM, the weather was bad. Since Bob was a very limited Private Pilot with no instrument flying experience, I had to fly that part of the trip from the right seat.

We eventually broke out north of Detroit and found the weather to be just a little bit better than predicted.

"That's a real break for us," Bob said, as we made our way

up the Michigan Peninsula. After three hours of flying, we decided that Pellston would be a good place to stop for fuel. We cancelled the instrument flight plan by radio as we prepared to land at the Pellston Airport.

We landed at the airport and everything appeared quite normal. We taxied up to the gas pumps, and as we all got out of the Tri-Pacer, the lineman came up and asked, "What the hell kind of a ship are you guys flying?"

"Now, that's an odd question," said Bob, "just what do you think he meant by that?"

"He probably hasn't seen a Tri-Pacer before," I jokingly replied. Further questions by the lineman left all of us just a bit perplexed.

"Does your ship always taxi with the tail down and the nose off the ground?" he continued. "Boy, you guys ought to have your main bungees checked, your gear looks like it's doing a spread-eagle as you taxied into the gas pumps."

With that, Bob, being over six feet tall, reached up and grabbed the wing tip. He shook it up and down several times.

The little airplane suddenly jumped up and the main gear now assumed the original regular normal position.

"Looks like you guys had quite a load, the landing gear was really strained," said the lineman. "Are you all intending to stay for a couple of weeks?"

"Heavens, no," replied Bob. We're only here for this weekend. By the way," Bob continued, "Don't worry about the looks, the bungees are just weak and it happens all the time."

We were all a bit reassured by his brief explanation as we climbed back in for our final leg to Drummond.

Arriving on Drummond Island, we found Doug Elliott and his dad had yet to arrive with the trailer. So, Mac, the owner of the cabin and land, borrowed a truck from one of his old buddies he knew on the island.

Mac said, as he loaded up the truck, "How in Hell did we ever get all that gear into this little Tri-Pacer? I thought Doug and his dad brought most of our stuff from Columbus."

We completed unloading our cameras, bed rolls, cushions, etc. and planned to enjoy our stay at Drummond as we all worked on the cabin and actually got in a little fishing on the side. The weather was perfect.

Finally "Mac" said, "Since this is Saturday we all should take time out and go into town tonight."

"Town, what town, Mac?" I said. "All I saw as we flew over was a cross road and a local saloon."

"That's it," Mac replied. "Soonies . . . everyone goes to Soonies on Saturday night."

Early that evening we drove up to the local crossroads from the cabin and arrive at Soonies. It was a quaint place. It had a good restaurant with a very lively bar. While we were eating at the end that was reserved as the restaurant, we noticed four Chippewa Indians a few tables away from us. One of these Indians was a woman, rather stocky and not real pretty. Mac quickly advised us that she was one of the local characters around here. We soon found out she was better known in this area as "Six-Pack-Peggy."

"Say, Mac, how did she ever get a name like that?" I asked.

"Oh! she is a local character here in town and she got that name because that's all it takes," replied Mac.

"You've got to be kidding," I said.

"No, I'm not," Mac said with a dead panned serious look.

Meanwhile, Six-Pack got Bob's eye and started to smile and wink at him.

Incredible. She had no front teeth . . . UGH! What a sight, as she guzzled down another can of beer.

"She speaks very little English," Mac continued, "but she sure knows the word beer, or six-pack."

Now, as our dinner conversation continued, two Indian braves with Six-Pack, started to argue in very loud voices. One of them was rather small, about 5 foot 7 inches. The other guy was almost half again as big as the little guy.

The language is in Chippewa or Soo, and we are unable to understand a single word that was said. "Probably best we don't," said Mac.

Mac, and the locals, are really enjoying the show, and it was free to boot. The noise and voices got more raucous, and I am beginning to get a bit uncomfortable.

I told the guys at our table, "I don't like the sound of this."

"Aw, there's nothing to worry about," assured Mac.

Just about that time there was a tremendous crash, and the sound of broken glass was heard quite distinctly.

Both Mac, Doug, Larry and I say, "It's time to get the hell out of here." Not Bob, he wanted to stay and see the fight. After all, Six-Pack is still making eyes at him and smiling.

Although we could not understand what they were saying, we did know, and it was quite evident, these two Indian braves were actually fighting over Six-Pack Peggy.

The little guy had the big advantage as he had hit his almost empty bottle on the edge of the table and was threatening to cut up the other Indian.

We made a hasty exit for the door, along with a lot of other patrons of Soonie's restaurant/tavern. Everyone that is, except Bob, and the fighting braves and Six-Pack Peggy. Bob insisted on staying and watching all the action. We accused him of having a longing for Six Pack. Of course he vehemently denied this.

Now the action really started as both of the Indians began throwing bottles, chairs, glasses, etc., anything they could lay their hands on.

It sounded and looked very much like one of those old time movies.

Those loud blood curdling screams! One right after the other, actually made our hair stand on end.

Finally, the bartender and another huge bouncer got the two Indians calmed down, and told them to go outside. Bob was still pretty much part of the show, trying to act as a referee, and we thought, ready to become part of the action.

Now outside of the bar, they appeared to be calmer, but the action and fighting broke out once again. Only this time it was on the dusty main street. Old Six-Pack Peggy was eating it up with all the attention that was being showered on her.

"Just think, they are actually fighting over her," I said, "that's hard to believe."

"Well," Mac said, "Fire-water does funny things to the whites, too."

The Indians had left now, and all was calm. We returned

back to the bar and restaurant. By now, our beer had gotten warm and our dinner was cold.

"So what," said Bob, who still hadn't gotten over the fact that this actually happened here on the Island. "Boy it's just like the movies, this will really be something to tell the folks at home about."

"Yeah, it's really something," Mac said, "that's why they don't like to see Indians drink. That fire-water gets them acting up every time. Of course, it does the same to us too, but not as fast as it happens to the Indians."

The next day was Sunday, so we packed up and left Drummond late that afternoon for Columbus, Ohio. We all felt that this trip would be remembered for a long, long time, and we got a lot more than we had bargained for. Our flight home turned out to be quite tranquil, and four tired guys were thankful to be home again safe and sound.

Chapter 12
"Cleared to Where?"

A good friend of mine, retired General Edisel Clark and I were together again on a trip with Ebco's Twin-engine Piper Aztec. This particular trip was scheduled from Columbus, Ohio to Indianapolis, Ind. The Task? Fly a group of Ebco's (Oasis Water Cooler) businessmen to Indianapolis, Indiana and return to Columbus late that afternoon.

We were scheduled to depart about 8 AM and the plans were to return sometime around 5 PM. The weather was bad, but flyable. Columbus weather was 400 ft. overcast with a mile and a half visibility. Indianapolis was a bit better with 1,000 ft. overcast and three miles in light rain.

The forecast called for improving conditions later on in the day.

I planned to meet Ed at the Ebco hangar about an hour before our scheduled take off time. It was good to see him again. It had been almost six months since our last contact. Both of us had a lot in common, we loved to fly.

We Filed our IFR (Instrument Flight Rules) flight Plan with Columbus. In the briefing, Columbus assured us that no ice had been reported in the last several hours along our path of flight. The only conditions we faced was the marginal weather

A twin-engine Piper that was similar to the one featured in this story.

in the Columbus and Dayton area. Indy reports showed that it was slowly improving to much better conditions.

We loaded up our passengers and were ready to depart on schedule.

It was nasty this morning, fog, drizzle, rain and no hope of getting on top. Some of the pilot reports showed tops to be at 15,000 plus feet. Not having oxygen we could not go this high.

There was the usual questions by our passengers prior to our departure. "How long before we reach Indy?" asked one passenger.

"About an hour and ten minutes," I replied. "Sit back, relax. We plan to serve coffee as soon as we level off, meanwhile you can read the local papers."

Ed and I hook up our head sets and were ready for contacting the tower. Our passengers could no longer hear any radio conversation.

"Columbus Tower, Piper Aztec N-2322-Yankee ready for our instrument clearance to Indianapolis.

"Roger, Piper Aztec N-2322-Yankee, you are now cleared to Indianapolis via the Plain City intersection, then as filed. Climb to and maintain 4,000 feet."

We did not have a transponder since this was prior to the days of radar and transponders. We read back the clearance and switched over to the Tower frequency for our take-off clearance.

We climbed out from the airport and made contact with the base of the clouds at 400 feet.

"Well that part of the forecast was right," I told Ed. "I only hope that the rest of the weather holds up to the weather briefer's forecast."

"Right on," replied Ed.

The tower requested we switch over to the departure control frequency. Now we established radio communication with departure control as we climbed up to our assigned altitude and reported level at 4,000 feet.

Departure control came back on the air with a further clearance. "You are now cleared to Indy via the Plain City intersection direct Dayton via Victor 50, direct Indy."

Reading back our clearance for confirmation, we tuned in to the Plain City intersection as directed.

Turning to the passengers, I informed them, "We are now at our assigned altitude and on our way to Indy," and I assured them, "our flight should be real smooth."

The rain by now was beating a steady tattoo on the windshield. The engines purred like two fine tuned sewing machines, and everything was just great.

Then without warning, it happened!

First one radio went dead, then slowly the second radio did the same thing. Next, the generators went out. "Unbelievable Ed!" I said. "They both just don't all die like this."

"You're right Mark." We both stared in disbelief as we watched our electric gyros go next. "According to flight rules, we are supposed to go directly to our last cleared fix." I say to Ed, "but just how the hell are we going to do that when we have no way to know where the Plain City fix is without our radios?"

"You're right Mark, so it really presents a major problem now," said Ed.

Both Ed and I, had our headsets on and there was no way our passengers could know what was going on.

Not wanting to alarm our passengers, we hesitated at this point in time to tell them about our present situation.

"Well Ed . . . I just don't know how we can possibly fly anywhere near the fix, much less even to try and go on to our final destination without radio navigation. Why, we can't even communicate with anyone, except ourselves."

"It looks grim, Mark," said Ed. "We don't have much choice but to try and go back to Port Columbus."

"Go back? Go back to where?" I said. "As we both know, the weather was very bad with a 400 foot ceiling and only 2 miles visibility, in rain and fog."

"All we can do, Mark, is make a 180 degree turn . . . let down to the minimum altitude on the airway. Then hope like hell the ceiling is still holding at 400 feet. Perhaps we might even break out and make ground contact if we're lucky."

"Whew! that's hoping for a lot. Remember? We do have

several large TV towers located just north of the Port Columbus Airport," I said, as we switched off our now dead radios. What do you think happened, Ed?"

"At this stage of the game Mark, I'm not sure. Do you have any ideas or thoughts at this time?"

"Nothing Ed, I only hope we see the ground before we hit any trees or towers," I replied.

"According to my calculations, we have been out about 12 minutes and now with our 180 turn, we should start down to our minimum enroute altitude," said Ed.

The time was fast approaching when we will be forced to leave the safety of this altitude and we could only hope and pray we would make ground contact and not get killed. "I imagine that Dayton and Columbus must really be getting worried about us not checking in Ed."

"That's the least of our worries now, Mark," replied Ed.

Ed was absolutely right. My mind raced feverishly as I tried to come up with some ideas.

"Say, Ed, before we start our final let-down to . . . maybe ground contact, why not try just once, to see if we might possibly have a spark of life in the generators. All we need is just one on the line. Maybe we might get lucky. Miracles still happen . . . occasionally."

"OK, Mark, but it's useless," he said as he reached over and turned on the #1 generator switch.

"Look Ed! . . . it's starting to show an indication. Maybe it's not completely dead after all."

"Let's try #2," I suggested. "What have we got to lose?"

"OK, here goes," said Ed as he switched on #2.

"I'll be damned, that just came back on the line too," said Ed with a big sigh and a slight scowl on his face.

"Let's try at least one of our radios . . . now," I added hesitantly.

Ed switched on the radio, and almost like magic, the one set came alive with a voice calling our aircraft number.

"Aztec 22-Yankee, this is Columbus radio, do you read?"

"Roger, we have you now, loud and clear," I replied.

"We have been trying to reach you for the past five minutes, request position report, Over."

"Roger, Columbus, we are just east of Plain City, and doing a 180 turn back to the intersection you cleared us to. We had complete radio failure, but now everything is back on the line and under control. We can proceed as cleared."

"Are you sure you want to go on, or do you want to return to Columbus?" asked Columbus.

At this point, both Ed and I felt we should go on. Everything was back to normal.

"We intend to proceed as cleared," I replied.

The rest of the trip to Indy was uneventful and we landed at Indy without any further problems.

However, we both were very concerned as to why we had complete radio failure.

The weather cleared as predicted for our route back home, and as a result, we were not too concerned about our return trip home.

"Well, with the weather completely VFR we can easily make it home for the radio shop to check it out," said Ed.

As per the schedule, our passengers began arriving and we made ready to go back to Columbus shortly before 5 PM.

Ed and I went out to the Aztec, loaded up our passengers, and prepared to start the engines.

Switches on! . . . Not a sign of life. "Strange," I said. "Do you think the battery is dead and that was our problem, Ed?"

"Could be, but now we need to get a battery cart for a power jump start," said Ed.

"Would you get out and go to the main office and see about getting some line guy to help us with a battery cart start?" asked Ed.

"I'll be happy to," I said, as I got out of the Aztec and went for help at the main desk.

I approached the front desk and the girl asked me, "Can I be of help to you?"

"It appears that our battery is dead and we will need a boost start," I replied.

"No problem," she said, "I'll call our line service."

Standing next to this girl was the owner and operator who

had heard my request for a power cart to start our engines.

He said, "What kind of an airplane are you flying, son?"

"A Piper twin-engine Aztec," I replied.

"That's no problem, I'll go out and prop you," he said as he made his way out the door ahead of me.

"But . . . but . . . but . . ., you don't have to do that sir, we're not in that big of a hurry," I replied.

"No problem son, hop in my jeep and I'll give you a ride out to your ship and help you get it started."

I arrived at the Aztec and told Ed, "I can't believe this Ed, that's Colonel Roscoe Turner. He has to be at least 70, and here he is, wanting to give us a prop?"

The Colonel had positioned himself on the right propellor and before he touched the propellor he yelled, "Switches on?"

"They're on, Colonel," I replied.

"Brakes on?"

"Yes sir, the brakes are on."

He swung the prop with all his might and managed to pull the propellor through a couple of turns. But, the engine failed to start.

"Say Ed, do you think we should let the Colonel do this, he is no young chicken, and I can hardly believe what I see."

"If it doesn't start by this next pull, we will get out and wait for the power cart," replied Ed.

"Boy, Ed, that Colonel Roscoe Turner is a feisty guy. I only hope that I can do that when I reach his age."

"Hang in their fellows. Let me give it another wack," yelled the Colonel.

With that next pull, the engine roared to life.

Now with one engine running we are able to use it to power start the left engine.

The Colonel smiling, salutes us.

"I can't believe that," I said to Ed. "We should be saluting him."

Our take off and flight home was in excellent weather conditions. We both were quite relaxed and prepared for anything. But, nothing of any consequences happened.

With no pressures on us, we had a chance to try and

analyze just why we happened to have all our radios fail us at the same time. We both agreed it had to be the ship's battery that gave us this problem. Everything else was operating OK.

"When we get back to Columbus, let's make sure either you or I contact Lloyd Young, and advise him of our problem," I tell Ed. Lloyd was Ebco's chief mechanic.

"You can bet on that Mark, he should be able to find out exactly what caused our complete radio and generator failures."

Lloyd tackled the problem the next day and turned up a very interesting answer. "Mark, I found out that the boss had decided to put in a Ni-Cad (Nickel-Cadmium) battery. These batteries are a new type battery that just came on the market. They are great batteries but unfortunately they have one big draw-back. When they are old, or lose energy, they almost always fail. But . . . they do come back to life just as fast."

That logically explained our problems, and we were advised that after our little episode, the Aztec's Ni-Cad batteries definitely will be exchanged for the regular lead cells. No one would ever want to get caught in a situation like that again. They might not have been so lucky!

Chapter 13
Single-Engine Night-Instrument

I had been contacted by a pilot friend of mine who flew for this small company to see if I could help him. He was planning to go on vacation and wondered if I could fly his company's single engine Beechcraft Bonanza during that particular time.

"I sure could," I replied, as we went to the hangar together to look over his ship.

He said, "My company doesn't fly a lot, but when they fly, they fly hard, meaning long days."

This was only to be a two week fill-in and the way he talked, I didn't expect to be flying very much.

I always did like the Beechcraft products, and the Bonanza was a fairly fast single engine airplane.

Later that week I was very surprised when I got this call requesting me to fly the Bonanza from Columbus, Ohio to Hartford, Conn. The trip called for an early departure from Columbus to Hartford and then depart early for the final leg to White Plains, New York, with a planned RON (Remain Overnight). The following day we would be flying back to Ohio with our departure time set for sometime after lunch.

"Boy, that's great," I said to my wife Ruth. "I'll have a

A Beechcraft Bonanza similar to the one pictured here provided this author with a night flight that will long be remembered.

chance to spend an evening with my family back in Rye, New York."

I met my passenger at the airport, a guy by the name of Dave Smith of Richardson/Smith Company.

We instantly hit it off. Dave was a big guy, extremely personable, and one couldn't help but like him. I filed the IFR (Instrument Flight Rules) flight plan for the first leg of our proposed trip. The weather wasn't too bad and we took off about 7 AM that morning. The flight was uneventful from Columbus to Hartford, Conn.

A Corporate pilot's flight time is minimal compared to the many hours of long wait time. This trip was no exception.

It was a long day and the evening and darkness had rapidly replaced the beautiful day. I did not look forward with any enthusiasm in anticipation of flying a single engine airplane at night, even though it was this nice single engine Bonanza.

I checked with the weather service station and was advised that the east coast weather had started to deteriorate.

"If you left now," said the Flight Service Station gal, you just might make the Westchester Airport, in White Plains, N.Y., before it gets too bad."

Well, that's not the way things turned out. I got a call from Dave about 5 PM advising, "Mark, we are running a little late. Looks like we should be out to the airport in about an hour and a half or two at the most."

"Understand Dave, I'll be ready," I replied, knowing that this delay would probably put us in some rough weather conditions. In addition to this, I did not feel comfortable flying a single engine at night in the New York area.

I updated my departure flight time with the FAA and at the same time asked, "Anything new on the weather in the Westchester area, and the New York area in general?"

"Sir, when do you expect to get there?" she asked.

"As an educated guess, looks like it will be about three hours from now," I replied.

"Well, you should be OK, but, they do forecast the weather to deteriorate after dark to bare instrument minimums of two hundred feet and a half mile in rain and fog."

As promised, Dave showed up at the airport a good two hours beyond what he told me. "How does the weather look in New York?" asked Dave.

"Not too good," I replied, "but I think we still should be OK by the time we get to White Plains."

"Let's do it," he said. "My meeting is set for tomorrow, so we don't have to rush."

I cranked up the Bonanza and checked over all the equipment, particularly the lights, to make sure everything functioned properly. I got the clearance and we were soon airborne and on our way.

The weather started getting bad about thirty miles south of Hartford. It started to get dark now, and the weather was not good. They advised me there could be the possibility of ice in the clouds below 4,000 ft. After this warning about ice, I elected to fly at the minimum enroute altitudes. It started getting real dark now and I had to turn on all the lights, both inside and outside of the ship.

Everything was going along very well, until I realized that no one was talking on the radio. I thought, damn, that's strange, particularly in this busy east coast area. I'd better give a call to the New York Center and see if they are still reading me.

"New York Center, this is Bonanza 39-Alpha calling, over." No answer, only this time I had a bit of urgency in my voice. "New York Center, this is Bonanza 39 Alpha calling, over." Still no response.

Then an unfamiliar voice came on the air. "Hello Bonanza, 39 Alpha . . . this is American Flight 454 calling, do you read?"

Now that's odd, I thought, just why would an airliner try to contact me?"

After establishing contact, the American pilot advised he was passing a clearance on to me from the New York Center. "They have been trying to reach you," he remarked "and we told them we had heard your transmissions."

"Roger, understand American, go ahead with the clearance."

"Bonanza 39 Alpha you are now cleared to the Westchester Airport, via direct the Radio Beacon, for an ILS (Instrument

Landing System) approach to runway 14, maintain four thousand."

"Understand, American, thanks," I said, as I read back my new clearance limits to him.

"New York, this is American . . ." I didn't hear New York's reply and I wondered why American's radio conversation stopped.

The radio came back on with, "OK, New York, the Bonanza has acknowledged the new clearance and I advised him to contact the White Plains approach control, 20 miles north."

"OK, Bonanza N-39-A, New York has gotten the message."

"Thanks a lot for the help American."

"No problem little feller, have a good trip," replied the unknown friendly American pilot.

Now it is completely dark, and rain started beating on the windshield. I'm in clouds with moderate rain and personally looking forward to landing soon at the White Plains airport.

Finally, I'm now in range of the approach controller. Boy, it's good to talk to a ground controller again after so long a period of silence.

"Bonanza 39-A, this is Westchester Approach Control, you are now cleared to the Radio Beacon for an ILS 14 approach."

What a comforting feeling, as I called approach with a request. "Westchester approach, how far do you show me from the Outer Marker? (radio beacon)

"I don't know sir, you're not in positive radar contact yet. Report the marker."

"Roger," I replied.

I had my approach charts out and all set up for the ILS instrument approach. Dave heard my comments to the various controllers but, not being a pilot, didn't say much to me.

"Dave, I had better get this gal slowed up so I can get down," I said, as I lowered ten degrees of flap.

Strange, was it my imagination, or did the lights go dim, as I put down the flaps?

It was about this time when Westchester Approach came

on the air again and advised me, "We now have radar contact. We show you a mile from the maker and number two for the ILS approach. Contact the County Tower at the marker."

I had the ADF needle locked on the outer marker and began to pick up the localizer and glide slope indicators for the ILS/GS (Instrument Landing System/Glide Slope) approach.

"It was time now for the landing gear and check list," I said to Dave, as I reached out and put the landing gear handle in the down position.

Then, without warning, all the lights just about go out.

I immediately checked and found both the ammeter and voltmeter indicators showing a full discharge.

I can't believe it, one minute I'm ready to intercept the final approach course and the next minute all my radios and lights are acting crazy. The radios are absolutely dead. The light in the cabin is so dim I can't even tell if they're on.

"What's wrong!" I asked aloud. "What's wrong!"

It was about this time when I think I hear a faint radio call from the Tower calling me.

"Bonanza 39-A are you having any problems?"

My God, what a silly question, of course I am.

"I just lost the ILS signal," I replied.

Damn, I wonder what Dave must be thinking about now?

"Understand, advise you go back up to 4,000 feet and hold west of the VOR [VOR, a navigation system]."

Before I get an opportunity to reply, I broke out and could visually see the city of White Plains all lit up in a myriad of colored lights.

I got on the radio and immediately advised the tower that I have made visual contact with the ground and requested, a contact approach to the Westchester, White Plains Airport.

"Sorry Sir, but we are not sure of your position at this time, suggest you maintain special VFR (visual flight rules) and fly east until you are out over Long Island Sound. Hold southeast VFR and we will try to work you in on a special VFR."

He continued by saying, "You may expect approach clearance in about 35 to 40 minutes. We have another instrument approach ahead of you, and you will have to wait your turn."

The visibility had slipped down now to about two miles. I gratefully acknowledged the clearance and raised the gear and flaps to reduce drag.

Now I understand! The lights dim again and I lost the navigation signal and radios, just like I did before.

"It just has to be our generator," I tell Dave, who at this point is just anxious to get back down on the ground. I'm happy that he doesn't seem too upset by this turn of events. I found the coastline and set up a holding pattern between Port Chester and New Rochelle, New York, and flying at the same time 1,000 feet above the water and ground.

I used the back course needle to show me each time I passed the end of the runway.

The Westchester Tower advised me, that I may have to hold here for at least another half hour to forty-five minutes.

"Due . . . to the increase in the in-bound traffic," was his reply.

I started to reduce all of the unnecessary power drain that put the load on my electrical system. I quickly found that for a change, the radios were beginning to return to normal.

"Say, Dave," has this ever happened to you before with this ship?"

"Well, to be honest with you, Mark. Yes, it has," said Dave. "Sure is a fine time to find out about it," I say to him.

Apparently he hadn't heard my remarks as I received no further acknowledgement from him.

As we go back and forth along the coast for the umpteenth time, the weather began to deteriorate. We are now down to about 400 feet with one mile visibility. I'm getting concerned and I could tell Dave must sense this.

I only hope we can get our clearance soon. I'm not sure just how much longer I can continue holding in this deteriorating weather. With this under powered electrical system acting the way it is, I'm not too sure if I can rely on my radios to safely get me back on the instrument approach system.

Finally, the tower called and advised me, "You're cleared for the special VFR approach to the airport."

It's absolutely pitch dark, pouring down rain and getting

very foggy. I gratefully accept this special type of clearance and VFR approach to the airport.

From out of the darkness this voice with a nasty smart ass remark said, "A special VFR clearance? That guy is nuts."

After the forty minute hold, I didn't really care what that guy had to say. I can't say that I feel this was the time to discuss my problems. I only wanted to get back down on the ground.

Finally, the airport is in sight and we land safely in a pouring down rain storm. It was still pouring as Dave and I took off in different directions.

"Mark, see you tomorrow at 10 AM, so long," was Dave's parting remarks.

The next day dawned bright and beautiful. All of the foul stinking weather had cleared out. I met Dave at the airport and as we approached the Beechcraft Bonanza, we both whistled. The "V" tail Bonanza used to have a beautiful paint job, but now it really looked weird. All of the leading edges, wings, and the "V" tail were silver. There was absolutely no paint left as a result of flying in all that heavy rain last night.

As we flew back to Columbus, I talked to Dave and asked him, "Do you ever remember the radios acting up like that before?"

"Come to think about it, why yes we have, Mark. But we were never flying at night with all of the equipment in use at one time," he replied.

After landing at Don Scott-OSU Airport, Columbus, Ohio, I checked the ship over thoroughly and found it only had a very small 25 Amp generator. Far too small to handle the electrical loads that were required.

My parting words to Dave were, "My God, Dave! We really could have busted our butts with that inadequate generator. I would strongly suggest you change from this 25 amp to a 50 amp generator before anyone even thinks about night flying again."

"Ahhhh . . . you're absolutely right, Mark. I'll be sure to take care of that, later.

About a month had passed when I found out the company had taken care of this problem. They bought a twin-engine Cessna pressurized airplane.

Chapter 14
The Test Pilot

The company I was working and flying for was known to be quite aggressive. We had been flying a Cessna Skylane for company business during the past year. The boss, after checking over the facts, decided that now would be the perfect time to step up to a higher performance airplane. Perhaps it was the novelty or maybe the potential publicity, whatever it was, he decided that the new Cessna 210 would be THE airplane for his company.

I was thrilled to think we had finally gotten a good airplane with much better speed and a retractable gear. This would be the very first Cessna 210 in our area. Being the first of its kind, it really did create its share of publicity. The gear retraction was different and caused many a head to turn every time we took off.

It was a fast ship. I was able to flight plan it at 190 MPH, and it actually did what the books said. This company airplane was a true business airplane. My boss used it to make a multitude of trips throughout Ohio and some of the surrounding states.

One day in late fall, Les Rausch approached me and said, "Do you think we can fly up to St. John, Nova Scotia in one day?"

Pictured here is the later model Cessna 210. A great single engine airplane. Unfortunately, the earlier model described had a few problems that needed to be worked out before it would ever reach this status.

"I don't see why not," I replied. "It is within flying range. How about letting me check it out, and get back to you?"

I took my maps into the back room and laid out the most direct course that would take me to St. John, in Nova Scotia.

"Les," I said, "it looks like it will take about four and a half to five flying hours with one gas stop."

"That's great, Mark, plan on a trip to St. John's for next week-end."

That's how our new ship was initiated on its first long cross-country flight. As planned, we departed for St. John that week-end. I wondered why he had picked a place like St. John's to fly to.

I soon found out, as we departed Columbus Les said, "I'm looking to buy Campobello Island. Have you ever heard of this place?"

"Les, isn't that the late President Roosevelt's summer home retreat area?"

"Right on, Mark," Les shot back.

Somehow or other, Les had found some extra money that he wanted to use to invest in property. His intention was to find something that was different, unique, and wouldn't lose its value.

On our way over, I planned a gas stop at Portland, Maine. I particularly wanted to check the weather for St. John, N.B., and the surrounding area.

"Well," I said to Les. "The meteorologist said . . . it looks like we can make it. But . . . they are expecting some weather to move in about the time we get there."

"In fact, I was advised they expected it to start snowing just about the time we were scheduled to arrive, and it probably would snow all night."

Filing my flight plan, we charged off for St. John's. It was as predicted. The weather started turning sour just about 70 miles out of St. Johns. Here we were, out over the Atlantic Ocean in a single engine. This in itself was not too comforting, but fortunately the steady roar of the reliable Continental 260 horse powered engine kept its steady beat. This flight helped bring back memories of the many long hours of single engine flying between Africa and Italy over the Mediterranean Sea during World War II. But, the Med was almost always warm most of the time.

The Atlantic was not very friendly this time of the year and the water temperatures in the area were extremely frigid. The chances of water survival would be rather slim at best. Several of my passengers queried me on this, but I assured them there was nothing to worry about. Despite all my assurances, they nervously played cards without really knowing or caring who was doing what, as they watched the churning Atlantic waters below us.

I contacted St. John approach control and was snapped back to stark reality when he reported that the weather was starting to close in. I pushed the throttle farther forward, urging my airborne steed to go even faster than it ever had gone before.

Another urgent call from the St. John Controller advised, "I say old chap you had better hurry! The weather is rapidly closing!"

I said out loud to Les, "Just how the hell can we hurry any faster? I'm at full power now."

I called the St. John tower and said, "It's no use, I can't possibly go any faster." It was just about this time I spotted the edge of the airdrome.

I called and advised them of my position, and reported I had visual contact on the field. "Roger, old chap! We do not have you, but you are cleared to land on any runway.

After I landed I started to taxi back to the parking area. We were now in a full blown snowstorm. It took some time, but I finally found the right taxi way to the local fixed base operator. He welcomed us as his long lost friends. Believe me, we were as happy to see him, as he was to see us.

Our Canadian host, Bill Van Horne was there to greet us.

Prior to meeting Van Horne, I asked Les, "Les, are you really sure you want to buy property in this cold snowy area?"

"Oh, what's a little snow?" he said, as we boarded Bill's big van for the trip to Campobello. Our host was a real estate salesman. Maybe it was just a pure coincidence, but, driving to the retreat of the late President Roosevelt, Van Horne said, "There is actually no connection between me and an Island I would like to show you, and it is much closer than Campobello."

"Oh?" said Les. I think that's all that Bill needed as he continued. "Incidentally, has anyone here ever heard about an island called, Van Horne Island? Its near St. Andrew by-the-Sea?"

Everyone of us in the van shook our head and at the same time Les said, "Never heard of this place."

"Well, believe it or not, it is a much better place than Campobello," replied Bill Van Horne.

Our group by now had consumed several bottles of that good old Canadian Champagne, and Les our leader, agreed he would be willing to take a look at this little island Bill called, Van Horne Island, located near St. Andrew by the Sea.

Van Horne Island was indeed a very interesting place. Every six hours the island became a peninsula. A car could actually be driven over the gravel bar and on to the island at low tide. At high tide, it was covered by fourteen feet of water. This area was close to the Bay of Fundy with its forty eight foot high tides. The whole area was quite impressive.

We had been drinking beer, along with champagne and scotch. As a result, we were all in a good mood. Les, in this very jovial mood, tried to write in the two foot snow bank, "I love you, Maxine," as he relieved some of the excess fluids he had been drinking for the past few hours. Unfortunately, he just didn't have quite enough left to dot the I.

The natives thought we were all crazy Americans. They were ready for bed and we were just getting started.

What everyone failed to realize was the five hours difference in local time due to the crossing of the time zones.

I can testify that the Lobsters were absolutely the best in the world. The unique Island/Peninsula location, the huge historic main house and the St. Andrew golf course patterned after the famed St. Andrews in Scotland, helped Les decide. He bought the island.

This was the start of our many trips from Ohio to the Van Horne Island in New Brunswick. I found an airport located in St. Croix, Maine, that was a lot closer to St. Andrew than St. John. It was almost a ritual as we were kidded by the locals when they said, "Remember now, this country is great, but there is only eight weeks of summer and the rest is winter." They were absolutely right, but that didn't matter to Les.

Starting in May, we flew up once a month until October. I got to know just about every possible route from Columbus, Ohio, to St. Croix, Maine. Secretly, I enjoyed every trip we made, as long as I was able to change the routes.

It was along about July, when we were once again on our now familiar trip to Van Horne Island, St. Andrews. I had deliberately flown this long leg from Rochester, N.Y. to Portland, Maine. We were at 9,000 feet in the sun and everyone was sleeping, except me. We were just about over the roughest part of New Hampshire when suddenly, without any warning, the engine quit! It was sudden silence.

I quickly moved the gas selector switch and switched fuel tanks. The engine started running again but very rough. All of my passengers were now wide awake and very nervous. "Can't be carb ice," I told Les. "This is a fuel injected engine."

I called Maine Center and advised them of my situation. "Looks like we might have to make a forced landing in this small river bed below us," I told Center.

"Roger, understand," they replied. "We're unable to get a positive fix on you, as you are slightly out of range."

Rechecking my map, I advised Center of my position.

"We're 30 miles NW of Concord and my engine is still running rough." My passengers must realize we are in a rather desperate situation.

I tried to reassure them that, "This ship is rugged, and we probably will be OK, even if we had to land in the river bed."

That little explanation did not help one bit.

We were down now to a little over 3,000 ft. I said to them, "Aw, what the hell have I got to lose?"

I switched the mag to the left magneto. The engine quit completely. Absolute dead silence!

I quickly switched back to both, the engine started to run again just as rough as before. Well, here goes the other mag. This is it, I say silently to myself. Now the engine began to run a lot smoother.

"That has got to be the problem," I said, as I called Center for a direct route to Portland.

I landed at Portland and the mechanics immediately found the problem and replaced the bad mag. We continued on to St. Andrew a couple of hours later and without any further problems.

On a trip to Harrisburg, Penna., we stayed overnight. When we returned to the airport the next morning, the fixed-base operator queried me, "Is this a new type of Cessna 210?"

"Why no, what do you ask?" I said.

"This is the very first time I have ever seen a Cessna 210 with a fully feathered prop," he said, pointing out to my ship.

As I looked out the window, sure enough, I can see the prop is in the full feather position. I found out later a by-pass valve had stuck in the propellor system which in turn had created a very unique condition.

On yet another flight, our indicated speed suddenly went from a cruise of 190 MPH to 150 MPH for no apparent reason.

Later, we found the hydraulic system had leaked out all the pressure automatically. Then slowly and silently, the gear doors opened and our gear dropped while we were in flight.

Another flight over New York, our speed once again started slowing up. "Damn, it can't be the gear, I just had that dumb thing fixed," I moaned to Les. I carefully checked the flap system and found a leaking flap valve, which in turn lowered my flaps.

"God, what next?" I said to Les.

We had a good week-end at the Island this month and decided we would depart early Monday morning due to the high winds we had on our nose. I had planned to land at Elmira, N.Y., for fuel.

After landing, I taxied up to the local operator. A young kid met us and I requested, "Please fill it up, and check my oil at 9 quarts."

"Sure thing," he replied.

"Boy, Les, this kid is on the ball."

"Is there a restaurant on the field?" asked Les.

"No, just a snack machine," he replied.

We slowly ambled over to the machine for a candy bar and a cold drink. A few minutes later this kid came rushing up to me all out of breath. He asked, "Did you guys have an accident with your ship?"

"Hell no!" I replied. "Not that I know of."

"Well, I suggest you had better come out and take a look. Your airplane looks awful."

"What in the world is he talking about, Mark?" asked Les.

"Damned if I know," I said, as we both go out to look at the 210. We arrived at the ship and I am shocked at what I saw. Along the top of both wings the metal had sunk down between each rib. Indeed, it looked like we had been in one hell-of-a-wreck. I didn't know what to say except, "Les, I'll be right back. I'm going to call the Cessna factory and try to find out what to do."

I reached Cessna and was lucky enough to talk to an old friend of mine, Jerry Dutton, the factory rep for my area.

He began by saying "By the way, Mark, what color is your ship?"

"Just what the hell does this have to do with my ship?" I replied rather indignantly.

"What color is your ship?" he said again impatiently.

"It's a dark chocolate brown, maroon and tan," I replied, adding, "with white trim."

"Tell me, are your wings the dark color?"

"Yeah, Jerry, they sure are."

"Were you flying at high altitude?" was his next question.

"Yes, we were at 10,000 feet," I replied.

"Well, yours is not the first, nor will it be the last. The answer is real simple. What happened was the temperature differential. It was real cold, and now hot. The cold air actually contracted the metal. Of course, the color had a lot to do with it. The cold air and the hot sun all helped create this problem. Actually only the top bay of each section collapsed. Right?"

"That's right, Jerry. Sounds like you have the answers, but it's truly incredible. Can I fly it home, safely?" I asked.

"Sure thing, Mark. Give me a call when you arrive, OK?"

Unknown to me at this time, the boss had taken his fingers

and actually pushed on the leading edge of the wing at each bay section. This in turn, oil-canned the metal back to its original position.

I flew home at a much lower altitude and at a much slower speed and we arrived home safe and sound.

As my boss exited the Cessna 210, he forcefully said, "Mark, that does it! There will be no more test-pilot flights. . . . Tomorrow we will buy a twin engine airplane."

Chapter 15
Hail . . . Hail . . . The Gang's All Shook Up

It was early fall and I had this trip posted for four passengers from Columbus, Ohio to Chicago, Illinois and return. This trip was scheduled for a 6 AM departure with intentions to return late that evening.

Two of the passengers were very good friends of mine, John Conville, and Ed Philpot. Ed was a former Pro-football player for the Boston Patriots. He was not too enthused about flying, but he just had to get to Chicago for that very important meeting today.

I had advised them there was a good possibility of marginal weather on our return flight that evening. I had hoped this would encourage them to decide to schedule their departure a little earlier than planned.

John was the first to arrive at the Meigs airport. Apparently he had heeded my warning about the storms that were developing.

"How's the weather, Mark?"

"Looks like we will have to make a detour, John, the weather around Ft. Wayne stinks."

"What does that mean in flying time Mark?" asked John.

"Well, the way I see it, we will go due south of Chicago to Indianapolis, then due east to Columbus via Dayton. Which

This Cessna twin-engine 310-Q is similar to the Cessna that encountered that ferocious thunderstorm. It had to be strong to take what it did. Photo courtesy of Bill Walker & Associates.

means another half hour to 40 minutes more flying time."

By this time the rest of the crew had arrived and were tuned in to my every word and a loud moan met this statement.

"I'm sorry fellows, but there are a lot of thunderstorms in and around the Fort Wayne area. Not having airborne radar in this ship has forced me to make this decision. Believe me, I'm not about to go through any storms. It looks like I can make this end run around Indy and then on to Columbus. Thus we will avoid most of the heavy weather." I felt speaking in football terms, Ed would understand my thinking.

Even with this simple explanation, they still wanted me to go direct. "Damn Mark, won't this give us a lot of extra flying time?" asked John. "Why not go direct? We will save ourselves a long tiresome flight that way."

I finally convinced them that this was the only way, and they might as well accept it.

I loaded up the twin-engine Cessna 310 and radioed Meigs Tower for the instrument flight clearance back to Columbus via Indy.

There was no delay for a change and they shot back my clearance as requested.

The flight back was very smooth and very little weather. There was a time or two when we did see a few clouds and as a result of this, I would get a smart-ass wisecrack about a pilot wanting to tour the countryside to build up his flying time.

However, we did see some very high CB's (thunderstorms) off to our east in the vicinity of the Fort Wayne area. I tried to explain to them, "Hey guys, look at those huge thunderstorms off to our right and in the vicinity of Ft. Wayne. Aren't you glad I chose to fly this route, now?"

"Yeah, sure Savage, I understand, just another damn good reason why you selected this route to build up more hours," replied Conville.

Indianapolis finally came in view and I turned the corner in order to fly eastbound towards Columbus. Everyone in the ship, except Conville, agreed after seeing all the large CB's off to the east of our track, finally admitted that they believed I did make the right decision after all.

We were in light rain and on instruments now, but the air was very calm.

It was quite natural for them to inquire, "Hey, Savage, are we there yet?"

"No," I replied. "Please just relax guys, believe me, the hardest part of this trip was the distance we had to travel to go around those storms. Dayton is just ahead and we should be home in about 35-40 minutes."

I called Dayton Radio and requested an update on the current weather system. "Dayton Radio, this is Twin Cessna N-3281-A . . . could you give me a report on the new Columbus weather?"

"Roger," shot back Dayton. "Columbus is 3,000 over-cast with five miles. The rain ended 10 past the hour. The forecast calls for improving conditions."

I had the loudspeaker on and turned to the gang and said, "Well, did you guys hear that? Looks like this was the right thing to do after all." Everyone nodded in agreement.

Despite the fact we were still flying in rain and poor visibility, I could sense that everyone had begun to relax.

About four miles east of Dayton, I called Dayton Radar once again.

Dayton replied, "Cessna 81 Alpha, be advised that the only thing I can see on my radar scope is a very small cell about 10 miles to the east of your present position. In fact, according to my radar, you should be out of the precipitation within the next 20 or 30 miles."

Now that was real music to my ears, as I flew on towards our destination. About five minutes later, I called Dayton once again for additional reassurance. "Anything new on that cell in front of me?"

"Negative, but if you want, you can detour either north or south of your present course."

"That's OK, I'll just fly on a little further and take a look," I replied.

About another two or three miles, and it started getting dark. Really dark. The air was calm, but the rain had begun to increase in intensity. Then, without warning, it got black. So black, that the day suddenly became night. A voice in the back jokingly said: "Hey Mark, who turned off the damn lights?"

I had to turn on my night lights so I could see the instrument panel.

The rain now had started to come down in buckets. It was the heaviest rain I had seen since flying over the great Amazon jungle during WW II. The ship began to shudder and suddenly started going up like an elevator.

I had control, but was unable to keep the ship from going up into a tremendous rate of climb. I pulled the power off and made an immediate left, 90 degree turn.

The rain was beating a steady staccato on the front windshields. Occasionally I thought I saw what looked like little white balls mixed in with the heavy rain. The noise was deafening.

Ed was shouting wildly at me. Something that sounded like, "Please, please. Let's get the hell out of here!"

I'm much too busy to even reply. The altimeter was winding up like we were in a jet. I finally got it stopped at 12,000 feet. I decided that the first chance I got, I would try and call Dayton radar. The airplane was like a wild animal. It was all I could do to keep it right side up.

I did have a chance to grab my heavy Jepp book and place it up on the windshield in front of my face. I was very concerned that the windshield might disintegrate from the heavy beating it was taking.

Ed was still screaming at me, but there was absolutely no way anyone could be heard over that awesome racket. Now on a 360 degree heading, all I could do was hope and pray that we would get out of this monstrous storm alive.

Suddenly, the ship found the back side of this shaft of uprising air. Now, full power, and we are going down, down.

"My God, I hope that this ship hangs together," I prayed aloud, knowing full well that no one could hear a word I was saying.

Back at our assigned altitude of 5,000 feet, I was finally able to regain some control, when all of a sudden, we were thrust up, up, and again almost out of control.

The heavy rain let up for a second, and I heard John moan then plead. Now screaming, "Let's get the hell out of here!"

"What do you think I have been doing," I shouted back.

The Cessna was in a steep turn to the left, this fast elevator ride did not last quite so long and I was able to get the ship under control around 9,000 feet.

Each time I reached for the mike, it would be jerked out of my hand by the tremendous strong "g" forces working on my body. Finally, at long last, I was able to get the ship back down to my assigned altitude of 5,000 feet. We were still in extreme heavy rain, but I could feel that the extreme strong up and down drafts had diminished slightly.

Now, for the very first time since the darkness settled in on us, our flight conditions seemed to have shown some type of improvement.

I picked up the mike for the umpteenth time and was finally able to get a call out to Dayton.

"Dayton, this is Cessna 81-A." I'm sure they could tell what kind of ride I was getting by the very way my voice was raising and falling with each transmission I made.

Before I could receive a reply, the dark night turned into dull daylight.

The rain was still coming down in buckets, but, at least we were able to see the end of our wing-tips. Ed, who was seated on my right, suddenly started punching me on my right arm, and at the same time started yelling, "Hey, Mark, LOOK! LOOK! We lost our right wing tip tank!"

I tried to remain as calm as I could, as I looked over to my right wing tip and said, "We sure as hell have."

"Well," cried Ed. "What are you going to do about it?"

"I'm sure as hell not going out and fix it now," I said, as calmly as I could.

That little remark didn't do a thing to calm him, or the rest of the passengers one bit. Dayton had come back on the air and requested, "Cessna 81-Alpha, what is your position?"

I replied, "We were in a very heavy thunderstorm and also encountered hail, and I couldn't contact you until now."

"Understand," was his simple reply.

"We are now about ten miles east of Mechanicsburg, with possible damage," I replied.

The light had improved rapidly now and we are able to look over the 310 much easier. The extreme heavy rains and the turbulence were gone.

The air was smooth and we only had just a light drizzle. We were still in instrument conditions with our flight visibility of less than a mile.

I could see where the right wing tip tank had disappeared from the tip of the tank back to the first bulkhead. It looked weird not seeing the usual wing tip-tank and running lights.

The left wing tip tank tip was in shambles and just sat out there dangling. I had no idea as to how badly damaged the rest of the ship was. But, we were still flying.

I contacted Dayton once again. Only this time I advised them of our present condition.

"Are you declaring an emergency?" was their excited reply.

Negative, we are still flying, but have no idea as to how much damage we have," adding, "I'm planning to fly at our minimum speed until we reach Columbus."

"Understand," replied Dayton.

My passengers, still in shock, were finally able to mutter a few words. They were not ashamed to freely admit, "Damn, it all, we thought it was the very end."

"The worst is over," I assured them, as we finally broke out into reasonably good weather.

Dayton had notified Columbus of our predicament and on my first contact with them, I was greeted with, "Is everything all right? Was anyone injured? Are you sure you can make it to the field? Do you need emergency equipment standing by? How badly are you damaged?"

I began to believe that they never would run out of questions.

I assured them that: "YES! we definitely did get hail damaged. That, NO! No one was injured, and I am requesting the emergency equipment, YES! I'll be glad to get back down on the ground."

As I landed at Port Columbus I could see all of the tower personnel with their field glasses trained on my ship. I taxied up to our parking area and deplaned all of my completely shook-up passengers. Several loud whistles pierced the air, as we all looked in awe at the battle damaged Twin-Engine Cessna 310.

The left wing tank tip was just dangling in the breeze. The right wing tip was gone. From the first bay forward, there was absolutely nothing left. It looked very weird without a wing tip, but it looked worse as we checked further.

All of the lights were destroyed, including the rotating beacons. Large dents were quite visible in the leading edges of the wings, horizontal, tail and nose.

Several of the guys moaned, "WOW! It's a miracle that we made it. I hope that we never have to go through anything like that again."

I said a silent prayer in thanksgiving and then said to all my passengers, "So do I!"

Chapter 16
A Night to Remember

It was late fall when I received a call from Paul K. Jr., my boss. "Mark, we need you to fly me and a few of our salesmen tomorrow from Columbus, Ohio to Long Island, New York. I believe the field is called Republic Airport, or some name like that. We would like to depart Columbus around 7 AM. We should be returning sometime around 5 PM the same day."

"No problem, Paul. But the forecasts call for weather to move in that area sometime early evening and this could very well delay our departure," I replied. Paul shrugged this off saying, "I doubt if it will delay us."

The next day at 7 AM I had Paul and the five passengers aboard the Piper Navajo. The flight to New York was routine, and we arrived at 9:30 AM. As Paul departed, he advised, "We should be back by three. Plan to file your flight plan for that departure time, OK?"

"Roger, understand," I replied. It was beautiful all day but by late afternoon as promised, the weather had started turning sour. By 4 PM, the rain and thunderstorms had begun pelting the entire New York area. No real problem. I had radar, boots and the whole nine yards.

The Piper Navajo after its conversion to the Panther. This ship as the original Navajo was involved in that night flight from New York to Columbus, Ohio.

Five o'clock came and went, still no Paul. Finally, at 7:30 PM, he showed up with his people.

"All set to go?" he asked.

"I'm ready when you are," I replied.

"Before we leave, how about a bite to eat," said Paul.

"Why not," I said sarcastically, what's another hour or so, delay."

I refiled my flight plan for the new departure time as we gulped down a hot dog sandwich and coffee.

Now, the weather was really pounding the area. Hard rain and winds buffeted us as we made our way out to our Piper twin-engine type Navajo. It had five seats in the cabin with a pilot and co-pilot seat for a total of seven. A real business type airplane. By the time we had reached the ship we were completely soaked by the heavy rains that were blowing through the area. The heat from our bodies created a fog that steamed up all of the windows in the ship.

I called the Republic Tower for my instrument clearance to Columbus, Ohio, and was advised, "Due to the weather, we have a traffic problem, and a possible delay of at least one hour, to an hour and a half, prior to your departure."

I advised my passengers about this delay and it was agreed they would rather leave the airplane for the warm and dry coffee shop, rather than wait in the damp airplane. With this decision, everyone got out of the airplane and made a mad dash in the heavy downpour back to the coffee shop. It was only natural that everyone got completely soaked to the skin once again. The wet rain created a chill and we all welcomed the warm coffee shop with open arms.

Paul said, "Tell me, Mark, is this weather any better as we go further west and back into Ohio?"

"No it's not, Paul. In fact this heavy rain changes to snow around the Pittsburgh, Pa., area," I explained.

"Will we be able to make it back to Ohio?" asked Paul.

"I don't see any problem, except for the winds. It looks like we will be bucking about a 50-60 Knot headwind at my requested altitude. So . . . it will probably take us a lot longer than the usual two and a half hours to get home." What an understatement that turned out to be.

New York had one hell of a reputation for its rapid-fire speech and delivery on radio. I was not surprised when the controller started talking. He helped continue this wretched reputation the minute he started talking. Without stretching anyone's imagination, he actually sounded like a tobacco auctioneer with his staccato instructions.

I climbed the loaded Navajo out of the New York area in heavy rain and headed for Ohio. Everyone was soaked to the bone, but felt relief as I finally topped out of the weather conditions at 10,000 feet. The stars were out and it was clear and beautiful, as we flew above most of the rain clouds and crappy weather below.

My boss liked to ride in the back, but tonight for some strange reason elected to ride up front in the co-pilot's seat next to me. There had been nothing to see for over an hour except the top of the clouds and the twinkling stars. The heavy thunderstorms had moved on east.

As we droned on towards Ohio, everyone fell asleep in less than a half hour. They all had a hard day in New York and were now trying their best to catch up on a little rest.

The trip was quite uneventful and as I mentioned earlier, our normal flight time was usually 2 and a half hours at the most. Now, it was three hours later when my boss woke up and said, "Are we near Columbus yet?"

I hated to inform him of our position and estimated time to Columbus, but I had no choice. "No Paul, we are just now passing Johnstown, Pa. and I'm afraid we will have to land at Pittsburgh for fuel."

"You've got to be kidding," said Paul.

"No, I'm not. We have had one hell of a headwind and I will definitely have to stop for fuel at Pittsburgh." I replied, adding, "Believe me, I'm not looking forward to this particular stop. For your information, the weather there is 200 foot ceiling with one mile in snow and blowing snow. In fact they are having a blizzard type snowstorm, the last time I checked."

"That's hard to believe," said Paul. "After all that rain in New York, and so clear up here."

"The real fact is, the cold front is still to the west of us and this help produce all that heavy rain in New York," I explained. "Now, the cold air has turned all that rain into a huge snow area. I hate to think about leaving this beautiful night with all the stars, and letting down into Pittsburgh. But we have to, I'm getting low on fuel."

"Can't we find a better place to land, other than Pittsburgh?" asked Paul. "Won't it be kinda tough sticking our nose in all that snow and junky weather?"

"You're absolutely right, Paul. But I have to land there, since I have no other choice. We can't make Columbus unless I get fuel."

He tried to explain this to his men, but was not very successful. They felt we should have been there by now and were very surprised to hear we were planning to land at the Allegheny County Airport in Pittsburgh, PA.

I called the local controller and was cleared to land, using the current instrument approach system known as an ILS (Instrument Landing System).

Boy, it's really blowing and snowing like hell in Pittsburgh," I tell my passengers, "but we will only be here long enough to get fuel." My comments were met by loud moans from just about everyone in the ship.

"My plans are to depart as fast as possible, then climb up to 10,000 feet in order to fly back in the starlit skies," I assured them.

It was snowing very heavy during my approach to the airport. The snow was coming down just as heavy after my landing, this in turn made it difficult to taxi back to the ramp and parking area for fueling.

I hated to get out of the ship, it was snowing that bad. The FAA rules required everyone to de-plane while we refueled. This was indeed a welcome break for my weary passengers.

I checked over the Navajo and was surprised to find a lot of ice had accumulated over the Navajo's wings and tail surfaces.

I took my flashlight and carefully inspected the leading edge de-icer wing boot surface. "Oh! Lord!" I exclaimed out loud. But, there was no one around to hear me.

I examined the boots closer and I detected a multitude of small pinholes in the de-ice boots. This had allowed all the water from the heavy rains in New York to collect in the boots. Now in the extremely cold air, the boots had turned to a solid ice build-up.

"Aha . . . that's why my boots were not working too well," I said to Paul, who was by my side now and acting as my supervisor. I had to remove this ice if we expected to take off again.

The only thing I had available was my pilot's clipboard. I started trying to chip off the ice. It was a long and laborious job. I decided I needed help and solicited my passenger's hands and fists. With our frozen hands we were soon able to beat off most of the accumulated ice that had covered the leading edge of the wings.

More than one of my passengers complained about the rough torture treatment I had assigned them to.

"OK, gang, hop aboard," I said. "Looks like we can be on our way to Columbus."

"Thank God for that," exclaimed Terry Nation. "I was beginning to wonder what kind of chores you planned to assign us next."

"Now, Terry. Let's not bother Mark, he has more important things to do to get us home," said Paul Jr.

"Can we continue our flight with the wing boots in this condition, Mark?" asked Terry.

"I believe so, Terry," I said. "We should be able to climb back up and get on top of all this bad weather before we take on too much ice."

"How much longer will we have to fly before we get home?" asked Ron Meese.

"I would say, probably about two more hours."

"TWO MORE HOURS! we should have been there two hours ago," moaned Terry.

I replied, "I know, I know, we could have stayed in all that weather and make it in much less time but, since the de-ice boots are not working, I had no choice but to fly high out of the weather and bucking the strong headwinds."

I thought I had made a very convincing argument, but I

realized that to a tired crew, nothing I could say would help them feel any better, nor get them home any sooner.

"The winds aloft are still very strong and directly on our nose," I reported.

"OK, Savage, you win. We will all gladly settle for a smooth ride on top," said Paul Jr.

After receiving my clearance and take off instructions, I poured on full power and clawed my way up and out of the raging blizzard below.

We were soon back up in the clear at 10,000 feet. I called and checked in with Pittsburgh departure control as per their request.

Back on top the stars are shining brightly and everyone has once again settled back for that final two hour flight home.

Normally I can fly to Columbus from Pittsburgh in about fifty minutes. I started to reduce my climb power setting from climb power to cruise, using the throttles and prop controls.

But . . . God God! I can't move any of the levers, they were frozen solid in the climb position.

Engines and props were winding up close to their red lines, as I fought to keep the ship from climbing. I honestly debated about returning to Pittsburgh and back to the raging blizzard.

"Paul," I said. "Would you like to go back to Pittsburgh in that awful blizzard, down through all that ice, and snow?"

"Absolutely, no way!" cried Paul Jr.

OK, I agree. I'll strike out for Columbus," I replied.

With that decision made, I leaned over to Paul Jr., and said, "If worse comes to worse, I can probably land at Wheeling, St. Clarsville, or even Zanesville."

The engines are moaning and groaning in their present frozen condition. Operating at 10,000 feet with frozen propellor governors speeding up, then slowing down, does not make for a comfortable relaxed ride.

Without warning, first the right engine surges. Then about the time it settles down, the left one does a repeat performance. The ship sways violently to the left and now to the right.

I called Pittsburgh Departure. "Pittsburgh, this is Piper Navajo N-34-WC, for your information, I have a problem with frozen engine controls."

"Is this an emergency? Do you request a steer back to either Allegheny County or Greater Pitt?" requested the controller.

"Negative," I replied. "We are holding our own, and I plan to head towards Columbus."

With my de-icer boots not working properly, I was not too willing to go back down in all the ice and heavy weather again.

"Roger, understand. If your situation changes, or you need assistance, we will have a clearance for you, just advise."

We are in the clear, yet I was very uneasy about this new development. Fortunately on my departure and climb out, I did not pick up too much ice on our boots as they are now completely inoperative.

I checked Pittsburgh for the latest Columbus weather and was advised, "The storm had passed Columbus a few hours ago and the Columbus weather was not too bad. They were reporting, "light snow, but good visibilities. Twenty five hundred feet overcast, with five to seven miles in light snow."

Out of the corner of my eyes, I can see Paul had this frozen look on his face. His arms were folded across his chest indian style. He had not even blinked an eye for some time. His eyes were riveted on my every move, as if he was trying to figure out what I might say, or do next.

God, I thought, he just has to say something soon. He hasn't moved a muscle in over an hour. He just sits there staring at me all the time.

I can't take it any longer as I looked over towards Paul, I put my thumbs in both ears, fingers spread apart and wiggling, I yelled, BOO . . .! That little maneuver didn't even phase him one bit.

Finally, about ten minutes later, between engine surges, he managed to say, "Are you sure we can make it?"

Not wishing to alarm him, I said, "Hell, yes. I wouldn't be flying this now, if I thought we couldn't."

That statement seemed to quell his fears, for now.

I was turned over to Indianapolis radar by Pittsburgh and after I established contact with them, Indianapolis radar said,

"I understand you have a problem? Has your situation changed yet?"

"Negative, we still have a problem, but we are holding our own now," I replied.

Each time our engine surged, I thought for sure we would throw off our prop.

Paul, after more than two hours of complete silence finally started to talk. "Mark, I am of the opinion that if we got in trouble, you would still be smiling as we hit the ground."

"You're absolutely right, Paul," I replied and let it go at that.

I am now with Columbus Approach Control, and they asked me the same questions as the other controllers.

I advised them, "we are at 10,000 feet and request this specific altitude until I am positive that we can make the field."

"Roger, understand," shot back his reply.

Without my mixture controls to shut off my engines, I had envisioned shutting both engines off with either the gas control or the mag switches, in order to land. His next question shook up the troops.

"Do you want the emergency equipment to stand by?"

"Roger," I replied, as I started my let down through all the clouds. Both engines were still doing that weird surging.

"Now," I said half aloud and without thinking, "I hope that this crate holds together until I land."

"What's that you said?" asked Paul Jr.

"Nothing . . . really," not wanting to get him all upset again.

Finally, the runway loomed in sight. I got ready to cut off both engines just in case I can't reduce the power with the throttles.

"One more time," I said to Paul, as I reached for the controls hoping that a miracle would happen.

"What happens if you can't pull the mixture controls back to shut off the engines?" Paul said.

"I'll just turn the gas handles to the off position on both engines," I replied.

"Ready? Now . . ." I say, as I reached for the controls once more to see if they will move. It's like a miracle, my prayers have been answered.

The throttle, props and mixtures moved, and they moved freely. I can reduce the power for the first time since we took off from Pittsburgh, more than two hours ago.

It had been more than five hours from the time we had left New York. Yes, indeed. It was a night to be remembered . . . forever!

Chapter 17
Hey Boss! . . . I Can't Let Down

I departed out of Columbus, Ohio with our company airplane called a turbo-prop Beechcraft King-Air for a trip with my boss to Toledo, Ohio. We finally topped out the bad weather at 10,000 feet. This was another business trip for my company and, for a change, I had a co-pilot with me. His name was Bradley Troy.

Brad was young, but quite aggressive. He loved to fly and had just gotten his pilot's license. I always welcomed the opportunity to take someone with me who was this eager, as it gave me the opportunity to teach and share some of my pilot time and knowledge with them.

I had filed an IFR (Instrument Flight Plan) from Columbus to Toledo, Ohio. My boss expected to be in Toledo for no more than a couple of hours, and then we were to go on to Detroit. Our intentions were to arrive back in Columbus sometime after six that evening.

"This trip should be easy, as we only had a couple of stops to make," I told Brad. He was a bit disappointed as he had heard about some of my other trips where we would land and take off at least 10 or 12 times during the day.

We had gotten our instrument clearance from Columbus

King-Air N-5115-Delta as it appears in its normal landing configuration.

and after breaking out on top, I set the auto-pilot and left Brad up front to watch the office.

He was quite willing and I felt no hesitation about assigning him this particular responsibility.

Before leaving the front cockpit, I told Brad to take my captain's seat as soon as I left. He said to me, "No sweat Boss, if I need anything, I'll just shout."

I told Brad, "Be sure you leave the overhead speakers on, I want to be able to hear any new instructions that the controller might give us."

I went back to the rear cabin and sat down to talk to my boss. This was one of those rare moments I had to take this opportunity to have a meeting with him and not be disturbed.

Brad felt, and acted like the 'King of the Hill,' which was fine by me. It always gave me a lot of satisfaction when I was helping someone like him.

Even though I was in the back, I still kept an ear out for any radio calls that had my aircraft number mentioned. While I was talking to Paul Jr., my boss, I heard Toledo Approach call, "King-Air N-5115-D!"

Brad returned the call almost immediately.

"King-Air N-5115-D, you are cleared to descend to 4,000 feet, you can expect an ILS runway 25 approach."

Brad repeated the clearance correctly and glanced back at me for my approval.

"No doubt about it, Paul. Brad does have the makings of a good corporate pilot."

"Gee, I wonder why he has not left 10,000 feet?" I asked Paul, aloud.

"Don't know," replied Paul. Of course Paul, not being a pilot, didn't understand, but was quite willing to agree with whatever I had to say.

"King-Air 15-D, please advise when you plan to start down for the approach," requested Toledo.

I started to get up and see what was wrong with Brad, when he turned his head back towards me and said, "Hey, Boss, I can't let down!"

I honestly felt that Brad was playing a practical joke, and I just waved and said, "OK, Brad, let's take it down, NOW!"

Brad replied in an almost pleading way, "I'm serious, Mark, I can't let down."

I hurried back up to the front cockpit to see just what the hell was going on. Brad got out of my Captain's chair and I took his place.

"Brad, are you sure you disengaged the auto-pilot?" I asked.

"Hell, yes, that's when I found out that nothing would work," replied Brad.

"What do you mean, it wouldn't work?" I asked.

"The trim tab wheel and the elevator wheel won't move," replied Brad.

I released the auto-pilot control and tried to push forward on the large trim wheel, and the elevator controls and nothing happened.

"By God, Brad, you're right, it just won't budge an inch."

Now things were beginning to look a bit grim.

Next I tried to push the control wheel forward again to let down.

"I can hardly budge these controls," I tell Brad.

"Yeah, I know Mark, I have already checked that out and got the same results."

Toledo Approach came back on the air with: "King-Air N-15-D, I still show you at 10,000 feet, you were cleared to 4,000 feet for the approach."

Picking up the mike, I called Toledo and told them, "We have a problem!"

"What is the nature of your problem?" they shot back.

Not wanting to alarm my passengers, I had switched from speakers to the headsets when replying to Toledo Approach Control's call.

I replied, "I'm not sure, but it looks like we have a trim control problem."

"Understand," they said, "Keep us advised."

"Well, Brad," I said. "At least this will give us a bit of a breather while we can sort this problem out."

"Do you have any idea as to what the problem might be?"

asked Brad.

"Sure don't Brad. In fact, I've never encountered anything quite like this before."

"So . . . what are your plans, Mark?"

"At this point in time, I just don't know," I replied.

"Do you know something, Brad?"

"No, what's that, Mark?"

"I'm damn glad I have you along to help me."

"I have no idea what the problem is and I am afraid it could be real serious."

Both Brad and I went through the checklist again to make sure we did not overlook anything.

"As you can see Brad, we are in the clear now, at 10,000 feet, so how about calling Toledo and getting their weather."

"OK," he said, as he called Toledo as per my request.

"Toledo Approach, could you give us your current weather conditions?"

"Roger, King-Air 15-Delta, the current weather for Toledo now is, 400 feet overcast . . . two miles in light rain and fog."

"Well, that's great," I tell Brad. "This indeed will make for a very interesting flight."

I tried to move the trim wheel and control wheel again but got the same results. Nothing!

"It's no use Brad, it looks like we will have to try and bring this gal down together. I don't think I have enough strength in my arms to hold the required forward pressures for the whole approach."

Brad then asked, "Are you sure we can get this ship down safely?"

"Hell, Yes!" I reply confidently. But, way down deep I felt we are facing a very serious situation.

Experimenting, I tried pulling back on the power levers to see what would happen.

The ship slowed, but I found that the controls still took a considerable amount of forward pressure even to establish a 500 foot rate of descent.

"Call Toledo Approach Brad, and advise them that we're ready to make the approach," and I added, "tell them we are requesting the emergency equipment to please stand-by."

Brad was visibly nervous now and said, "Is it that bad, Mark?"

"I'm not planning on crashing, but we should not overlook anything, Brad. At this point, I'm not sure what's going to happen when we get on that glide slope and drop our gear and flaps for the landing."

"Understand," said Brad, as he picked up the mike and called Toledo Approach Control.

"Toledo Approach, this is King-Air 15-D, we are ready to descend and make the approach. Oh! by the way. We are requesting that you have the emergency equipment please stand by."

"Roger, understand," shot back Toledo.

"Well, Brad, this is it!" as I reduced power and started down into the clouds. "Stay with me on the controls, I'm sure I'll need all the help I can get."

"Roger, I'll do my best," Brad replied.

Now on solid instruments, we are committed. I felt quite confident we will be able to get down, but it's the fear of the unknown that creates the terror in everyone.

"A piece of cake," chirped Brad.

"Maybe, I said, but, we're not down yet."

We are passing through five thousand feet and both of us are beginning to feel the tremendous physical discomfort, as we push the control wheel full forward, trying to hold the ship in a standard 500 foot rate of descent. The King-Air fights both of us, as it tries to stay in level flight.

"Hang in there . . . Brad," I said. "I'm going to need all the strength you've got."

My arms by now are starting to ache from pushing the wheel forward against the strong locked elevator pressures.

"I'm doing the best I can to help you," moaned Brad. I could tell by his voice, he was getting fatigued.

It was just a short six months ago he had a very serious back operation. He was injured on the job when he responded to a medic emergency call. As a result, he had to medically retire from the fireman/medic job.

Brad, as an EMT (Emergency Medical Technician), found a woman unconscious under her dining room table. It appeared that she had suffered a heart attack. The woman was rather large, over 200 pounds. Brad, in attempting to pick her up, screwed up his back. Although he was fine now, I did not want him to re-injure his back pushing too hard.

Now, I am fervently praying that he can hang in there long enough with me, and hope his back won't give out.

As we passed through 3,000 feet, I called the Toledo Approach control and requested an up-date on their current weather.

"Roger," he shot back. "We now have 400 feet with one and one half miles visibility in rain, drizzle and fog."

He paused momentarily then said, "For your information, the emergency equipment is standing by."

"Well, that's comforting," I tell Brad.

I'm really getting tired now, and even though I have reduced the power to help let down, I still need the power on to maintain the rate of descent and help lessen the pressures needed on the control wheel.

The control wheel pressure fought both of us constantly. If we let up for just a second, the nose starts to pitch up immediately.

"King-Air 15-D, how are you doing now?" asked Toledo.

With both of us on the controls, I hesitated to answer, but used my one hand to press the mike button and replied, "Fine, Toledo, but we will feel much better as soon as we get this ship on the ground."

The outer marker light came on to signify that we are now 4.9 miles from the field and the airport runway. "Hang in there Brad, we're almost there."

By now, Brad's arms and mine are to the point where they are starting to shake with fatigue. I'm on instruments and still in the clouds. Brad is visually straining to see the ground and the runway.

Finally, the middle marker light blinks and the horn comes on, indicating we're only 1.2 miles from the end of the runway. Brad shouted, "Runway in sight! Runway in sight!"

Touching down we flashed by the waiting rescue equipment, and for the first time we both started breathing easier again.

I taxied back to the hangar and had a mechanic look over the King Air.

After a short inspection, the mechanic found the problem. A broken pulley wheel had caught the trim control cable and wrapped it bird-nest style all over the pulley and surrounding area.

"Just thank God, Brad, we were not in a climb, or we would have never made it."

"Yeah, mark, it does look like our prayers were heard after all."

Chapter 18
The Ice Machine

Today's flight with Industrial Machinery's King-Air was a scheduled trip that held promise for a heavy day of flying. The first leg of this proposed flight would be from Port Columbus International Airport with a direct flight to Buffalo, New York. Later that afternoon I was to fly over to Chicago, with a return trip back to Columbus for late that evening. The boss, Paul Jr., had me plan to depart out of Buffalo about 3 PM, for a direct flight to Chicago. We would then depart out of Chicago around 7 PM for Columbus and arrive sometime around 9-9:30 PM. Quite a full day.

As this was such a long trip, I decided to call around and see if I could find a pilot who wanted to fly with me.

Several calls later, I hit jackpot. Joe Holbrook our well-known local TV weatherman, agreed to go with me. He was a pilot with an instrument rating.

Years ago, in the early 50's I got Joe interested in flying and he took his first ride with me.

We both checked weather for this flight and found that the latter part of our trip might prove very interesting.

The Weather Bureau charts were forecasting some weather to develop during the latter part of the day in and around the Chicago area.

A front view showing the tremendous ice build-up on the nose, of the ship. It looked like it had been parked outside rather than just arriving at Pal-Waukee Airport in Chicago.

It was my usual practice to always inquire whether my copilots could stay over an extra day.

I asked Joe, "Would it create any problems for you and your evening weather TV show, if we don't make it back?"

Joe replied, "No, not one bit. I have taken care of that possibility by having a stand-by available, just in case."

"Great," I said. "But I am planning to get back to Columbus by 10 PM at the latest. This should give you ample time for your 11 PM TV weather show."

I had a full load of people, including Joe and myself, scheduled for Buffalo and Chicago, and eventually Columbus. Whenever we flew to the big cities, my boss always believed we should fly with all the seats full. Today was no exception, we had all nine seats full.

Joe had never flown in a King-Air before, and this trip would give him a real opportunity to see how it performed. In all probability we would get some real instrument flying time on this trip. Just what the doctor ordered for a TV weatherman.

It was a typical winter day, as we loaded up and proceeded to fly to Buffalo, New York. We had a high overcast with excellent visibilities. We also had a very strong tail-wind but no actual weather. "With this strong tail-wind, our actual flight time to Buffalo could be in record time," I remarked to Joe.

"It sure feels good to be back flying, Mark. Boy, this King-Air is some machine. Too bad we don't have some real weather to fly in."

"Don't worry Joe, as you know, according to the forecasts we are bound to get into some actual weather before we get home," I said.

After an uneventful trip we touched down in Buffalo. "Well now comes the hard part, Joe," I said.

"How's that, Mark?" asked Joe.

"The waiting," I replied.

Paul Jr., the boss, said, "Expect a one PM take-off time."

But, I knew better.

"Don't you believe it, Joe. Probably closer to 3 PM, would be more realistic."

As the time approached for our flight to Chicago, I thought

it would be a good idea to recheck weather and file the new instrument flight plan. I was not surprised when I was advised that a band of weather had developed just west of Chicago and was scheduled to be either over Chicago, or just east of the city, about the time of our arrival.

"Probably will start as rain and then change into snow later on," said the Buffalo weather service.

"Boy, that's great, Mark," said Joe. "This will give me a chance to fly actual instruments."

"Yeah, right, Joe," I replied.

The King-Air had a five hour fuel range with altitude, so I felt we would have more than ample fuel to make this leg of the trip with our remaining fuel. As fate would have it, this proved to be a very wise decision by not adding more fuel and the extra weight.

My boss showed up at 4 PM and we loaded up for the second leg of our flight.

"If everything goes as planned, we will be back in Columbus by 10 PM at the latest," I reassured Joe. "Should be ample time for your 11 PM weather show."

Joe did an excellent job as my co-pilot. He handled all of the radios and was a real pro. Having an instrument rating did help. Normally the trip from Buffalo to Chicago would take about two hours. With three and a half hours of fuel, I felt we would not have any problems. The legal requirements for instrument flight is you must have enough fuel to make your destination, plus your alternate airport, plus 45 minutes reserve.

"We meet more than the FAA requirements," I told Joe, as he enquired about the range of the King-Air.

As the FSS predicted, we started picking up clouds around the Battle Creek, Michigan area.

Our flight west-bound produced a strong head-wind that only a few hours ago, was welcomed as a tail-wind. I elected to fly at 10,000 feet, rather than at our normal altitude of 20,000 feet, because of the very strong winds. We are flying in solid clouds and I noticed we are beginning to pick up just a little ice on our wings.

"Say, Joe, how about calling Chicago and requesting if we

can get a higher altitude? No need to stay down here in all this ice."

"Roger," replied Joe.

I could tell Joe seemed a bit disappointed because I did not want to stay in actual instrument conditions.

We climbed up to 16,000 feet, our new assigned altitude, and were able to top out all of the clouds.

"That's OK, Joe," I said. "I'd rather fight the headwinds than sit down there in all that ice. We'll have plenty of actual weather flying before the day's done."

As it turned out, this proved to be a real understatement. I recalculated our new ground speed and found out to my surprise, how much more head-wind we had picked up. In fact, the new figures gave us both quite a shock.

"At this altitude, the wind has got to be really blowing, Joe. We're only doing 90 knots over the ground. This is a far cry from my usual 230 knot speed."

We're over Pullman, Michigan, and less than 85 miles away from Pal Waukee. The Pullman VOR is located on the eastern edge of Lake Michigan almost directly across from Chicago. Our radio is alive with all kinds of calls from Jumbo Jets complaining about the severe icing conditions over the Chicago area.

"Well, Joe. You got your wish after all. We should be getting into that tough weather very shortly when we start our letdown into the Chicago Pal-Waukee Airport."

"You know, Mark," remarked Joe. "It's hard to believe the weather is that bad below us. It's so nice up here above the clouds, with all the stars shining."

"Right, Joe. But, it has to be rough with all those big guys yelling about it."

Then our radio crackled loudly as we heard this Jumbo Jet, an American 747, give his pilot report of severe ice on climb out from O'Hare.

"Joe, I think we will try and stay up here until we get real close, then we will request an unrestricted descent direct to Pal-Waukee."

"You're the boss," Mark, said Joe, as he radioed this request to Chicago Approach Control.

"Roger," Chicago replied. "I'll see what I can do for you at this time."

"Understand," Joe replied.

"Well, Joe, this trip will be a lot longer than I had planned, but we should have enough fuel. I sure as hell would hate like hell to go into a holding pattern with all those actual reports of known icing."

Normally, the flight time between Pullman to Chicago was only 10 to 15 minutes. We have already flown 25 minutes and we're still 15 miles away at 16,000 feet and in the clear.

Chicago called and advised, "I plan to keep you at this altitude just as long as I can."

"Roger, I appreciate that," I replied.

"Boy, Joe, now that's music to my ears."

Joe had told me earlier that, "Even though he did have a current instrument rating, he rarely if ever, attempted to fly in weather like this, much less in actual icing."

Finally the time came when Chicago called and said, "Can't keep you up there any longer, need you to descend to 3,000 feet for the approach to Pal-Waukee."

"Roger," said Joe. "We're leaving 16 for 3 now."

We started down and entered cloud tops at 15,000 feet.

"From here on in Joe, I believe that things will get very interesting."

I pulled back the power levers and we began our long descent to our newly assigned altitude.

About 11,000 feet, we started picking up ice. Not just a little, but a lot. Out of the corner of my eye I could see Joe beginning to squirm around in his seat. Except for my small section of heated area, our windshields were completely covered over by this white opaque sheet of ice.

I must admit, anyone not accustomed to seeing this much ice, will definitely receive a very strong shock to their nervous system. It is pitch dark now, as I switched on my wing ice lights in order to see how much ice we might have picked up. Wow! my wings were loaded with ice. This didn't help Joe one bit after he saw that.

"Joe, would you feel better if I flipped on the de-icer boots?" I asked.

"You'd better believe it, Mark," said Joe in a bit of a nervous voice.

I left the ice lights on, and flipped the de-ice switch to activate the boots.

"It's like a miracle, Mark, the ice just popped right off," said Joe.

Joe sighed, and I could see him relax for the first time since we encountered all that ice on our wings.

"We're not down yet," I said, as we passed through 8,000 feet on our way to 3,000 feet. Despite the fact we're now in a rapid rate of descent, we're still picking up additional ice. I flipped on the de-ice lights once again and this confirmed my worst suspicions. The wings have as much or more ice than before.

Joe on seeing this, got extremely quiet and went into a very somber mood. I tried to comfort him saying, "Can't say as I blame you, Joe. It does look a bit rough right now."

"You can say that again," replied Joe in a very somber voice.

"Well Joe, you wanted to fly in real weather and ice, now you've got your wish."

"But . . . Yea Gods! I never dreamed it could be this bad, Mark," replied Joe.

My windshield heat managed to clear a hole in the ice, but it was a mighty small hole. Joe's windshield was covered over completely.

"Do you think we'll make it?" asked Joe.

"That's not the type of question that I have an answer for at this time," I said, trying not to alarm him. Down deep, I must admit I did have some real concern.

We're now level at 3,000 feet. The airspeed has dropped off to 140 knots IAS (Indicated Airspeed). The ice build-up on the wings and windshield was unreal.

"Boy, Joe, I'm glad we didn't fuel up in Buffalo."

"Why is that, Mark?" asked Joe.

"I don't think we would stand a ghost of a chance trying to carry all this extra load."

The King-Air N-5115-D as it appeared the next day. It had spent most of the morning in the warm hangar getting de-iced.

The ship staggered like a drunken sailor. "Damn, I hope that controller doesn't delay us on our approach," I tell Joe. "I don't see how we could even be able to do a missed approach if we had to."

I got the feeling that I'm talking to myself, as Joe just sat there not saying a word.

"15-Delta, you are now cleared for the approach to the Pal Waukee Airport, via the 15 ILS approach. Report to the Pal Waukee Tower, at the marker. For your information, the current weather at Pal Waukee is, 600 ft. overcast with one mile in light snow."

"Roger, thanks for the weather. We're clear for the approach," said Joe in confirmation, as he snapped back to life again.

"Joe, I'll let you know when to lower the gear. I don't expect to drop the gear until I am at the Middle Marker."

"The Middle Marker? . . . why that's less than a mile from the end of the runway," said Joe.

"Right, you are Joe. With this load of ice and passengers, I don't need any extra drag to slow us down."

The Outer Marker lights came on. Then shortly after the Middle Marker lights flashed on, Joe dropped the gear.

Joe was straining to see out through his ice covered window. He was barely able to make out the runway and the flashing lights.

"Runway in sight," yelled Joe.

I looked up, and at the same time saw the dim outline of the runway lights in the falling snow. Ah . . . what a great sight that was.

I was still holding my speed at an indicated 140 knots and decided I would not put my flaps down. I touched down at 140 knots and it was a feather landing. As I rolled out, I pulled back the power levers for full power in the reverse mood and brought the ship to a safe and sound stop. I did not want to use brakes as it was quite icy.

"Whew! . . . Well, we made it," I quietly said to Joe.

After unloading at the terminal, both Joe and I walked out to look over the ship. Joe let out a loud whistle, and said,

"Damn Mark, that is one hell of a load of ice."

"No kidding," I shot back.

"Do you mind if I get my camera out and take a few pictures?" asked Joe.

"Be my guest," I replied.

Paul Jr. in his usual hurried way said, "Plan to depart for Columbus sometime around 9-10 PM tonight, Mark."

"Sorry, Paul. But, that is out of the question."

"Just what are you talking about?" demanded Paul in a very loud voice.

"We are not going anywhere! I have over 8 inches of ice all over the airplane and I can't get deiced until tomorrow. Besides, we still have that severe icing and snow warning to contend with."

Reluctantly, Paul agreed, saying, "We will all be staying at the Hilton tonight."

I checked into operations and put in a request for fuel and de-icing for tomorrow.

While Joe and I were talking to the girl at the main desk at Preister Aviation, a guy rushed into the lobby area and yelled, "Jesus-kee . . .rist! Did you see that red and white King-Air parked on the ramp with all that ice on it? I have never in all my life, ever seen an airplane with that much ice on it, and still fly."

He was so convincing that just about everyone in the lobby went out to see what he was talking about.

He came up to us and said, "I think you guys ought to go out and take a look. It's truly incredible!"

I replied rather quietly, "No, I don't think that we want to go out there and look. We were the guys who flew it in!"

Chapter 19
"Think Positive"

Yesterday's all day rain ended. Now an early morning fog had settled in and the whole Ohio area was down to almost zero/zero. I was quite surprised when my boss called and said, "Get the plane out, and plan on going to Oil City, PA about 9 AM."

"Well, OK, but . . ." I was cut short with "No buts, I'll be there around 9 AM," and he hung up.

Paul Jr., was the boss, and after all, one is supposed to do what the boss says.

I arrived at Port Columbus about eight and requested Line Service to roll out my Beechcraft King-Air. "Boy, the weather is really bad," remarked Ed Komarek, Lane's chief mechanic, and troubleshooter. "Are you sure you want to go in this stuff?"

I replied, "Ed, I'm not planning to take this trip until it improves."

I returned to my office once again and called the FAA/FSS (Flight Service Station). I requested the current eight o'clock weather at Youngstown, Ohio. This would be the third time I called them since five this morning. They up-date the weather every hour, on the hour, twenty four hours a day.

Youngstown was the only station close enough to Oil City, Pa. (Actually the nearest airport to Oil City, was Franklin, Pa.)

Unfortunately, Franklin does not have a weather reporting station located there. As a result, I only had Youngstown's weather reports available to give me an idea as to what I might expect at Franklin, Pa. "For your information, sir, Franklin should have about the same weather conditions as Youngstown," replied the weather briefer. "This forecast should be good for ten this morning. Incidentally, your alternate airport would probably be Dayton. They are clearing up rather nicely."

"Could you tell me please, what does the current Youngstown weather show now?"

His immediate reply was, "The Youngstown weather is: Sky overcast with an indefinite ceiling, visibility Zero, with drizzle and fog. The RVR (runway visual range) is showing 2600 feet. It's just above minimums for the runway, sir."

"Well, I hope Youngstown stays above the minimums," I replied, as I filed my IFR (Instrument Flight Rules) flight plan. I felt in my own mind that I would probably not really leave until 10 AM because of the current low ceilings and visibilities. I planned to use Dayton as my alternate.

As usual, Paul Jr., arrived on a full run, and about out of breath said, "Are we ready to leave?"

"Well . . . not quite," I replied.

"What do you mean by that?" asked Paul.

"The weather is rotten Paul. Why, we can't even see the North American Rockwell Plant across the runway. The Weather Bureau was calling this stuff pea soup. It's that way all over the eastern part of the state." And I added, "It would be better if we wait a couple of hours."

"Nonsense, I can't wait that long," Paul said. "I have an appointment in Oil City at 10:30 this morning. We have to leave early to make that schedule."

"I say again, Paul, it's pretty bad and it would be better to delay an hour or so until this stuff burns off."

"Do you mean that this weather is going to get better, Mark?"

"Yes, it is Paul, but it will be a while yet."

"Now Mark, I want you to think positive, the weather will

be better when we get there, OK?"

"Right, Paul, I will think positive . . ."

"That's great Mark, that's all I needed to hear."

I took a deep breath, and said emphatically, "WE ARE NOT GOING . . .!

"Whaaat did you say?" asked Paul in a rather high-pitched voice.

I calmly replied, "I say again, we are not going."

"That's what I thought you said. Apparently you're not thinking positive. Is it possible that this stuff will be burned off, by the time we get there?" asked Paul.

"It's possible, but not probable," I replied.

By now, several other pilots are within ear-shot and just shook their heads, agreeing totally with my decision.

"I'm sorry Paul, but if we go now, I'll be wasting your time and money just circling in the holding pattern, waiting for the fog to lift before we can land."

"But, that's for me to decide Mark," Paul shot back.

"I am thinking positive Paul, that's why I think we should really wait an hour to see what it will look like then."

"Nonsense, I'll miss my meeting and I'm the guy paying the bills," said Paul in sort of an angry voice.

Finally, I said, "OK, Paul. We'll give it a try. But I'm sure we will wind up in a holding pattern for at least an hour."

"Apparently you haven't gotten the message yet," said Paul. "It will clear up by the time we get there, believe me."

"Paul, I don't think you completely understand," I said.

"Oh, I understand exactly. "You're just afraid to go."

This remark really pissed me off as I tried to contain my emotions by saying, "You know better than that, Paul."

Paul just stood there challenging me, and waiting.

"All well and good but . . . Aw Hell, let's go," I said. Everyone there shook their heads as we trooped on out to the King-Air parked on the ramp. No question about the dense fog. The King-Air was shrouded completely by the fog. We could just barely make out its silhouette in the fog and dim visibility.

I tuned the radio to the local ATIS (Automatic Terminal Information Service) and left the radio on the cabin speakers for

Paul to hear. The first words out of the radio speaker were, "Ceiling Zero, visibility Zero, runway visual range less than 1,000 feet."

"Did you hear that Paul? The weather is really stinko."

I called Columbus Clearance Delivery for my IFR clearance and this girl's voice replied. "King-Air N-5115-D are you aware of the current weather conditions?"

"Roger," I replied.

Her next message was, "Are you a 135 Operator?" (Air Taxi)

"No, I am not, and I'm ready for my clearance."

I could tell from her remarks she felt the current weather conditions were not too hot. Perhaps there was a possibility that I just might change my mind if she cast some doubt about my leaving at this particular time.

A long pause . . . finally she came back on the air with my clearance. "N-5115-D you're cleared to your destination airport, via radar vectors to Tiverton, then your flight planned route. Climb and maintain 3,000 feet, expect 15,000 feet 10 minutes after, Squawk 5244."

Reading back my clearance, I received her acknowledgement with the additional comment of, "STAND-BY!"

"Now what!" I moaned. We were parked at Lane's with both engines running and making quite a bit of noise.

Now a male voice came on the air with, "15-D . . . are you aware of the current weather conditions?"

"Roger," I replied, "and I'm ready for my taxi and take-off clearance."

"Understand," he said. "By the way, where do you want me to send the flowers?"

Damn, that little remark pissed me off and went over like a lead balloon. Keep your cool, Mark, I said to myself. You'll need to be calm flying into this weather.

I was barely able to inch my way along the taxi way due to the extreme low visibilities from the fog.

"I must be crazy to do this," I told myself half aloud. I wasn't too concerned as the boss was sitting way in the back cabin and couldn't hear me. But, at least I know the weather will be improving. Good God! The Boss has even got me thinking positive now.

I finally got lined up on the active runway and requested a clearance for take-off. I could only see 3 runway lights.

"Not too much visibility," I yelled back to Paul.

Finally, I received my take-off clearance and gave a big sigh of relief saying, "Roger, King-Air 5115-D, rolling."

I had no problem on take-off and we were soon at 3,000 feet on our way to our requested altitude. I thought, maybe just maybe, I might hit the top of the fog around 1,000 feet or so, and even be on top in the blue sky above. No such luck, we never did get out of the fog.

After about fifteen minutes of flying, I remarked to Paul, "The weather's not looking too good, Paul. This fog might not break up before noon."

"I'm sure you're wrong, Mark. It will be OK by the time we get there. Remember? Think positive! Incidentally, I'll just have enough time to get to my appointment, thanks to you."

"That's if . . . I can land at the Franklin Airport, Paul," I replied.

From here on, there was no more conversation, as Paul busied himself at the table with his office paper work.

I called Youngstown approach and request their local weather conditions.

"Roger," came the reply. "Youngstown has a special this hour, ceiling Zero, Sky obscure, Visibility Zero, Runway Visual Range is 2,400 feet."

I checked my approach plates and found this was exactly the published minimum requirements for the approach at Youngstown.

"Thanks Youngstown, for that weather report. I would like to shoot the published approach to Franklin, Pa."

"Understand 15-D. You are now cleared for the approach, we have no current weather reported for Franklin."

I set up my approach plates and was convinced this was a useless gesture with the weather so low at Youngstown. It would be a miracle if I would get the ceiling and visibility necessary for me to make the approach. Without a qualified tower control, Franklin had higher published minimums than what Youngstown had just reported as their current weather.

"Well, here goes nothing," I said. "At least I'll give it a try, but I'm not about to go below the published minimums."

The fog was so dense, I really did not expect to see the airport. I set up for the approach, flaps, gear etc., and rechecked my approach plates. I felt I had best keep the Youngstown approach plates out, and ready to use, just in case.

I flew down to the published minimums and failed to see the airport. I gave the King-Air full power and climbed back up and flew the missed approach procedures.

I called Youngstown approach and advised them of my present situation.

"Roger, we understand 15-D, what are your intentions now?" questioned Youngstown.

"Do you have any idea as to when they forecast this fog to burn off?" I asked.

"Negative. It's very dense, and it could hang on until noon," they replied.

"Rats, I don't think it would do a bit of good to hold until better weather," I said aloud.

Calling Youngstown again, I said, "Requesting clearance for an approach to your airport."

Youngstown came back with my clearance and vectored me to the final approach fix.

Meanwhile, Paul was getting restless and started to yell, "Damn Mark, are we there yet? I'm running late."

"I know Paul, relax. I'm setting up for our approach now," as I dropped my gear and flaps once again and called approach control hoping to hear that they had just a little bit better weather.

Approach came back on the air with, "Our current weather is ceiling Zero. Visibility Zero. Runway Visual Range is still holding at a steady 2,400 feet."

I checked my approach plates once again and could see that the airport was right at the minimums. Radar vectored me to the marker and the ILS (Instrument Landing System).

Now firmly established on the ILS, I started in for the landing. Needless to say, it's very foggy.

I just passed the middle marker light when the dim runway lights popped up. It was indeed, a sight for sore eyes.

I touched down and applied full reverse power to slow the King-Air down and to a safe stop.

Youngstown Tower said, "King-Air 15 Delta are you on the ground yet?" They're unable to see me due to the dense fog.

"Roger," I replied. "I'm stopped on the runway."

"When you are clear of the runway, and on the taxi-way, please advise. Due to our poor tower visibility, we're unable to see you. Request you advise us when you are parked on the terminal ramp."

"Roger," I acknowledged. I cautiously taxied the King-Air towards the Terminal via the taxi way. I finally found the ramp area and swung the King-Air around to park. With such poor visibility, it was just about impossible to read the airport's name on the side of the building.

"We're here, Paul," I announced shutting down the engines. I heard the door handle click and the piston sigh, as the door dropped down and open.

A long pause . . . and then . . . "This ain't Oil City, Pennsylvania, just where in the hell are we?"

"We're in Youngstown, and this is as far as I go. If you want to go any further you had better get a cab, or rent-a-car."

Paul by now, was so mad, he turned beet red. You could almost see the smoke pouring out of him. He stormed out of the ship and headed for the terminal and a taxi.

"Well, it's all over and I'll probably be fired," I mused, as I headed to the airport restaurant and probably my last meal with Industrial Machinery Co. and Paul K. Jr.

Later that morning, the weather turned out to be just beautiful.

It was about 2 PM when Paul Jr. made his appearance. "You know something Mark? I was so God-damn mad at you this morning I had all intentions of firing you when I got back."

"I know Paul."

"I want you to know one thing," he said in a very loud

voice. "You're the only son-of-a-bitch in my company that can tell me what to do, and get away with it."

"You know Paul, you're right, and if that situation should ever change, believe me, I will not be working for you."

Chapter 20
Hijacked! . . .*
Part I
"The Set-Up"

I was the Manager and Chief Pilot of the Aviation Department of Industrial Machinery Co. I had been with this company for 11 years and watched it grow from the leasing of a Cessna 310 twin to what we have now. A Beechcraft King-Air, A Piper Panther (Colemill conversion on a Navajo).

I taught the company how to use aircraft and they learned very fast. The company had grown from seven million dollars to over twenty-nine million in this short period of time.

I had to work, and work hard. Sometimes I worked for several months before I got a day off.

It was a challenge and I must admit, I did enjoy it. The company had two twin Engine airplanes and two full time pilots.

It was about this time that our company became acquainted with an individual known as David Hostelley.

Dave was a CPA (Certified Public Accountant) and had been known to be a promoter, as well as an accountant. Perhaps this was due to all of the many companies he knew through his business contacts.

He had made a special deal with my boss to do the company accounting in exchange for using the airplanes. A fair enough deal, I thought, and I got to know Dave quite well.

Pictured above is the Piper Panther N-34 WC (Whiskey Charlie) with the government exhibit stamp on the picture. This photo was used as evidence by the government in the two trials of the accused drug runners and kidnappers.

I had accumulated over twenty five hours flying time with Dave and his associate Anthony Magi during the past six months. I had flown him and Magi to Chicago, Pittsburgh, New York, Canada, etc., and almost bi-weekly round trips from Columbus to Cleveland and return.

Yes . . . Dave and I had gotten to know each other quite well. We were friends. I had Dave at my house for dinner on several occasions. He even went to Church with me and my family. I knew his wife and kids as well as he knew mine.

I thought he had a lot on the ball, but there were many times I thought he had just a little bit too much bull-shit.

About six months after my acquaintance with Dave, Paul Jr., my boss, called me into his office to tell me about a recent phone call he had received from Dave.

"Mark," he said. "Dave called me from Cleveland this morning and asked about using the airplane for a special trip."

"What did you tell him?" I asked.

"I didn't. I said that was your department, and you would be the one to tell him whether you would, or would not, make the trip."

"Well, Paul, you own the ships, I only fly and manage the department. I'll make this trip, but only if it is good for the company," I replied.

"That's what I thought you would say Mark, so you decide. I told Dave you would have to give him the answer, not me."

"Where does he want to go, do you know?" I asked.

"No, I don't know at this time," replied Paul.

"Well, OK. I'll wait until he calls me, and then I'll decide whether I will go or not."

"No, don't do that Mark, I told him that you will call him and talk to him," said Paul.

I called Dave the next day after I got back from my trip. He said, "I would like to use the plane for a special trip from Cleveland to Chicago, then on to Oklahoma City, Dallas, Ft. Worth and back, in three days."

"Dave, you're crazy, that's too much flying for three days," I replied.

"But, you've done this type of flying before," said Dave.

"You're right Dave but, there is a difference now. I do have a choice, this time."

Dave then said, "Mark, tell me, what would it take for you to make this trip?"

"Dave, about the only way I might consider doing it, would be if I could take a co-pilot with me."

"Well, I really don't want a co-pilot, but if that's what it will take, then OK," replied Dave.

That following week, as agreed, I had a co-pilot with me. His name was Bradley Troy. We took off with the Panther known as 34-Whiskey Charlie for Cleveland Lakefront Airport late that afternoon. Upon our landing at the Lakefront Airport, we were greeted by Dave Hostelley. He introduced us to some of his friends. Both Brad and I were a bit puzzled, as the first question asked by one guy was, "Are there any bugs on the ship?"

"Bugs? Hell no! My ship was cleaned only a few days ago.

"The only bugs I know of, might be on the leading edges of the wings."

Apparently he wasn't listening to me, as my words fell on deaf ears. One of the members of the party went into the cabin and checked out the ship with a small black box.

"She's clean," he reported back to the other two.

Brad and I soon put the names with the faces. Dave Hostelley, Anthony Maji, (Dave's partner), Ray Garcia, and Dick Rizzo with another individual we found out later was Rizzo's son.

Hostelly left the group and Anthony Magi took us in tow. We departed the Lakefront Airport in two cars and were driven to the Swingo Hotel located on Euclid Avenue in Cleveland. They all talked about how nice it was to have us here.

Anthony Magi was the spokesman for the group. I felt quite comfortable with him since he had flown with both Brad and me on more than one occasion. "OK, Anthony, what's this all about?" I asked.

He began by telling about this mission and what was going to happen. "We want you to work with us on this project. We will be working with the government. The Federal Government, flying several millions of dollars worth of securities for this secret mission," said Anthony. "In fact, Ray here is a former CIA man."

Ray then reached into his pocket, and produced some type of authentic identification to back up Anthony's statement.

Brad was invited to leave the room by Magi.

After Brad left, Anthony then said, "We know all about you, including your war record, family, etc., but we do not know a thing about this guy you call Bradley Troy."

"Well, I can vouch for him," I replied.

"Not good enough," said Anthony. "We can't take a chance like that and as a result we have another pilot who will be flying your airplane. You and Brad will remain here with Ray Garcia. He will take good care of you, and will be your security man."

Raising my voice, I say, "I can't do that."

Then Ray, the guy with the gun (a 38 Magnum) said, "We prefer it that way!"

Reaching into my pocket, I pulled out the keys for the Panther and give them to Anthony Magi.

Note: The full account of this bizarre event will soon be available in my new novel I've entitled, "HIJACKED!"

*All of the names and places are a matter of Federal Court records. Due to the nature of this case, I was unable to speak out, or write about this, until the final trial was held and the records made public.

Chapter 21
Hijacked! . . .
Part II
"The Reluctant Guests"

Brad had been sent word to return to our table, and when he returned we all settled down for dinner. We were about half way through the meal, when a bell hop came up to our table and asked if there was a Mr. Savage here. Quite startled, I replied, "I'm Mr. Savage."

"You have a phone call," he replied.

"Now, just who in the hell knows I am here?" I asked Anthony and his group. My question was met with complete silence as I got up from the table and was escorted by the bell hop to the pay phones.

"Hello, this is Mr. Savage," I said.

"Oh, Hi! This is Ed," said this unfamiliar voice.

"Ed! . . . Ed who?" I demanded.

"Just Ed!" was his firm reply. "I will be flying your airplane."

"Gee, that's nice, Ed. What do you know about a Navajo?"

"Oh," he paused. "I have enough flying time and feel quite comfortable in them," he replied.

He then asked, "All the paperwork is in the ship, right?"

"I believe you will find everything is in order," I replied.

"No problem then," he replied, adding, by the way, do you have two ADF's (Automatic Direction Finder) in your ship?"

"No, I don't, and incidentally, why do you need two ADF's? You're *not* actually planning to fly my ship out of the country, are you?"

"No, I'm not," he replied.

I pushed him a little harder saying, "But, why would you want two ADF's? Be advised, that if my ship is taken to Canada or Mexico I have to advise my insurance agent."

"Have no fear," he replied. "I was only asking, that's all."

"By the way," I said. "What do you know about a Panther?"

"Never heard of it," he replied.

"That's what the hell my ship is, and *you* want to fly it?" I shouted at him.

"Calm down," he said. "Just tell me all about this Panther. I'll take good care of it, and we should be back by this weekend." I returned to the dinner table and told Anthony and his friends about this call.

Both Brad and I had no questions in our minds now, we knew beyond a doubt, there was more than meets the eye with our little group of friends. (?)

We both felt they knew all along, about Ed.

Anthony Magi continued to brief us as to what we might expect while staying in Cleveland. "Ray, will be your security man while you guys stay here. You have nothing to worry about. He will be with you twenty-four hours a day."

"But, Anthony," I said. "We can't stay here, we didn't bring enough clothes. We only expected to be gone three days at the most."

"No big deal," he replied, as he reached into his pocket and pulled out a wad of bills big enough to choke a horse. Counting out five one hundred dollar bills, he handed them to me saying, "You and Bradley go out with Ray tomorrow and buy yourself some clothes."

After dinner, Ray Garcia, who looked and talked a lot like the Desi Arnaz of the TV series, accompanied both Brad and me up to our room. I started to open the door of our room and was stopped by Ray and cautioned, "Don't touch that door!"

"Just what the hell is going on Ray? Why can't I open my door?"

Bombing agony persists

By Ted Wendling and
Robin Yocum
Dispatch Staff Reporters

Two-year-old Andy Houlihan remembers the night his father nearly died as "The Big Bang."

Nine months later, still wracked with nightmares, he asks when the huge plywood-covered living room window will be replaced. He doesn't understand that it may be a long time before life in the Houlihan household returns to normal.

Thursday, Andy's father, Charles J. Houlihan, recalled his night of terror. He talked to a reporter after a federal grand jury returned a five-count indictment charging two former Columbus men and a reputed organized crime figure from the Youngstown area with the extortion bombing of his Fairfield County home.

IT WAS 3 a.m. last July 27 when Houlihan, 42, carrying a revolver and a bag of money from his vending machine business, entered his home at 6525 Winchester Rd. in Carroll, Ohio.

"I thought somebody was in the house," he said. "My wife (Linda) is an habitual person; she never goes to bed without closing the drapes and turning out the light, but two windows were open and so were the drapes.

"When I closed the drapes, that's when I heard this loud crack."

The "crack" was the explosion of a remote-controlled bomb that had been packed under the window sill of the Houlihans' living room.

It shattered the front of the house and blew Houlihan about 20 feet across the room.

AS HE writhed on the floor, his wife and Andy staggered from their bedrooms. Houlihan said he took "at least 15 minutes" to make sure they were safe and to tend to his wounds. He then picked up a .357 Magnum and walked into the living room to see a man walking toward the house.

He said he fired one shot and the man fled.

Houlihan said he later learned that the shot he fired was at Stanley Schmikla, 40, a convicted bank robber currently undergoing psychiatric testing at the Federal Correctional Institution at Springfield, Mo.

INDICTED THURSDAY with Schmikla were Orlando "Orlie" Carabbia, 56, a reputed Mafia figure who faces trial Wednesday in Cleveland on charges that he smuggled a ton of marijuana into the United States from Colombia; and Carl Thomas, 38, now serving a six-year prison term in Milan, Mich., for bank robbery.

Named as an unindicted co-conspirator was another former Columbus man, Kenneth Fields, 27, now serving a six-year term for bank robbery in Texarkana, Tex.

After being released from Lancaster-Fairfield Community Hospital, Houlihan his family lived in Florida under FBI supervision, returning home for Thanksgiving.

Charles J. Houlihan and his son, Andy. Dispatch photo by M

This newspaper clipping describes the night of the Big Bang! Houlihan his wife and son Andy, did survive.

"We can't take any chances, that's why," he replied.

Both Brad and I are shocked and a bit bewildered to say the least, as he visually checked the door and nervously peered down the hall, looking at all the doors and windows.

"Why it's just like in the movies," said Brad.

That night, Brad and I spent a very uncomfortable time in the room. I was afraid to talk or say anything. We both eye-balled the room looking for any hidden mikes, or cameras that might be tuned in to our every sound or move. We knew Ray was in the room next to ours, with his door open, and listening to our every word.

Next morning, Ray asked, "What do you guys want to do?"

"How about going out and getting some clothes," I said.

"Good idea," said Ray, as he led us out to the nearest Army/Navy Surplus store. As we entered this store, both Brad and I say, "Absolutely, no way! It reminds us too much like prison garb and we don't need that."

We left the store and selected a nice men's shop a couple of blocks away. We picked out clothes that should last us for at least a week.

"Well, now that that's taken care of, what do you want to do next, you guys have been to the ball game, movies, dinner, jogging, etc., what else would your little hearts desire?"

"Do you mean we have a choice, Ray?" I asked.

"Why you sure as hell do," he replied.

"OK, then. I would like to go home to Columbus," I replied.

"Me too," echoed Brad.

"That's an unreasonable request," shot back Ray, "I think I know what you guys really need. You're lonesome and you need companions."

"No way, Ray!" we say together.

That night at dinner in a nearby Greek restaurant we settled down for a nice meal. Brad by now, had me completely hooked on Cherrie Jubilee's. It was only a few minutes later when I spotted these two Bimbos making their way over to our table. "Say, Brad, those gals are coming over here."

"Naw, that's your imagination, Mark," replied Brad.

"Imagination hell!" I whispered to him, as both of the girls stop at our table.

"Fellows," said Ray, as he got up from the table. "I would like for you to meet your companions for the evening."

"Oh, Oh," whispered Brad in my ear. "You were right."

"This is Cookie number one, and this here is, Cookie number two."

I whispered to Brad, "I wonder what's going to happen to us next?"

Chapter 22
Hijacked! . . .
Part III
"The Interrogation"

That following Sunday, Ray took me out to the Cleveland Hopkins Airport in order for me to leave and go back to Columbus. I had told him previously that I had scheduled a week flying with Abbott Lab's Beechcraft King-Air as their back-up pilot and I wanted Brad to come with me.

"That's OK for you to leave Mark, but Brad has to stay," said Ray.

"But why?" I asked.

"He can wait for the Panther to return and can call you when it does," replied Ray.

Not wanting to argue the point, I felt that I had best leave while I can.

That following Thursday night I had a phone call from Brad. "Hi, Mark, I just god word from Ray that they want to let me go home, what do you think I should do?"

"They? Who is Ray referring to Brad, do you know?"

"Haven't the foggiest idea, Mark."

"Do you really mean you are free to go Brad?" I asked.

"That's what Ray said," repeated Brad.

"In that case Brad, get your ass out of there and come on home."

That following Monday I checked into Industrial's office and met with Paul Jr.

"Good Morning, Paul," I said trying to be as cheerful as I can.

"Hi, Mark! Did you have a nice trip?" asked Paul Jr.

"Well, it was interesting," I replied.

"Did you get the money from Dave Hostelley for the trip?" asked Paul.

"Ah . . . Ah . . . No, I didn't," I replied.

Paul said, "Well then, did they buy the airplane?"

"No . . . not exactly," I replied.

"Well Mark, if they didn't pay you for the trip, and they didn't buy the Panther, where is the airplane now?" Paul said with a bit more irritation in his voice.

"I'm really not sure where it is, Paul."

At this point, Paul exploded with, "YOU WHAT?"

After all the smoke cleared and his blood pressure came down, I proceeded to tell him about the horrible two week ordeal that both Brad and I had just gone through.

"That's incredible, and impossible to believe, Mark. Would you object if I called in the FBI?"

"Hell No, Paul! I would like to get some answers too."

I was told by Paul to call Brad and have him come down to our office in order for Paul to talk to both of us together. Brad finally showed up and for the next three hours, we were cross-examined by the full staff of Industrial Machinery Co.

We had about 12 people in the room at the time, and the questions came at us like rapid fire machine guns. "Do you guys really mean to stand there and tell us that you were Hijacked, and didn't do anything about it?" asked Bill.

"Jeeze, this jerk called Ray Garcia, had a loaded 38 Magnum and was with us twenty-four hours a day. We couldn't even go to the bathroom without his being there to keep us company," replied Brad.

"Did you guys have anything you wanted to drink?" asked Sherry.

"Sure did, Sherry. Brad started me on the Cherry-Jubilee kick and I had one every night."

"You've got to be kidding," said Sherry.

"Weren't you allowed to use a phone?" asked Donna.

"No . . . in fact, neither one of us was allowed to use the phones," I replied.

"Well, come to think of it, yes, we did use the phone once," added Brad. "But, that was only to call our wives, and this guy Ray was with us at all times."

Finally, I said, people, please. "We have had enough, give us a break. I have one hell of a headache and it is way past our lunch time."

Paul interrupted me by saying, "The boys do look tired. I'm sure you all have many more questions to ask them. Perhaps after lunch we can continue."

Brad and I both breathed a big sigh of relief, as we left the office and sought a place for lunch. "Damn Brad, that was the worst third degree I have evere been in. With twelve different people conjuring up questions, I thought that they never would let us go."

"You can say that again, Mark."

No doubt about it, we were both quite drained from the multitude of questions that had been shot at us over the past three hours.

"I hate to think what we will be going through when we get back, Mark."

"What do you mean by that, Brad."

"Have you forgotten? We are scheduled to meet the FBI next."

We arrived back at IMCO's office and are introduced to two men by Paul Jr. "Fellows these gentlemen are from the FBI, Mr. Chuck Tossi and his partner."

"Tossi? Tossi?" I said. "Mr. Tossi, we have been dealing with, Hostelley, Rizzo, Garcia, Magi, and now Tossi. Is Tossi Italian by any chance?"

"Why, yes it is," Chuck said. "But I'm one of the good guys. Now, would you mind stepping in this office, alone with me and my partner?"

I just don't know exactly what to say, or expect. It's almost like being in the movies, except this is for real. The door slam-

The remains of my Piper Navajo-Panther N-34 WC (Whiskey Charley) as it lay in its grave in Columbia, South America.

med shut behind me, and now I am alone with the two FBI men.

"Have a seat, Mr. Savage. We understand that you have been through quite an ordeal."

"You can say that again, Sir!"

"Oh, you don't have to Sir us, we want to help you, and try to get all the information about this little adventure, you and Brad were on."

I shot back, "Adventure . . .? I don't think it was an adventure, I think it was a nightmare."

"We understand, so, why don't you start at the beginning."

I must admit, I was treated very fairly, and they seemed to bend over backwards to help me in any way that they could. They gave me ample time to think and answer all of the questions that were asked. At no time were they rude or indifferent to my personal feelings. None of my story was questioned in any way, whatsoever.

"I imagine you must be tired and would probably like to take a break," said Chuck. Rick hardly said a word during all this time. He appeared to me to be just an observer.

"Yes, I would," I replied.

"OK, Mark, how about getting your partner in here now for us?" requested Chuck in his soft voice.

The door slammed shut behind Brad and I couldn't help but hear loud voices coming from the room.

"Maybe they're giving him the third degree," Paul said.

It had been just about two and a half hours when Brad finally emerged with the two FBI men. I can't believe it. Chuck has his one arm around Brad, and the three of them are joking and smiling. "Well, what happened, Brad?" I asked.

"The first two hours were pure hell. They really gave me a cross examination like I was a criminal. Then, it was like night and day difference, they believed me, after both our stories checked out. Now, as you can see, we are good friends."

It took about five years for us to find out what really happened to IMCO's Piper Panther, twin engine airplane.

Two pilots flew it down to Columbia, South America and picked up a load of "Coke" (Cocaine) estimated to be worth over 50 million dollars. They took off from a dirt strip and hit a tree and crashed. Both pilots died.

The final disposition of the case did not get settled until nine years later.

(The complete "Hijacked!" story, will be featured in a full-length novel soon to be published.)

Chapter 23
Deer! . . . Deer! . . . What Deer?

It was early fall when my company scheduled a flight to Beckley, West Virginia to pick up a load of customers and return them to Columbus, Ohio. I was to fly back to Beckley later that evening after dinner. It turned out to be a very beautiful fall day for flying and the weather forecasts for a change, promised a very pleasant fall night.

My son, Rick, a Major in the Air Force, had been home on leave for the holidays. I had promised him I would try to get him on one of my flights if I had the room.

Today was the day. He would also have the opportunity to ride in the co-pilot's seat of the King-Air. He had told me all about his job with the Air Force as a Weather Officer. During this past year, he had been working with the Astronauts in the space shuttle program. He was quite proud of the fact he had participated in the last seven space shots.

He was well aware of my job as a Corporate pilot, and was willing to wait with me if it were necessary. In the earlier years I had taught him to fly gliders. He was a pretty good pilot and achieved his Private Pilot's license in just one summer. Perhaps it was through this type of flying he had dreamed of becoming an Air Force Pilot like his dad.

Unfortunately, his vision kept him out of the pilot's seat and he had to settle for next best as an Air Force Meteorologist.

It had been a long day, but, a phone call advised me that our guests were planning to depart the Columbus airport for Beckley sometime after dinner.

From my past experiences with our out-of-town guests, this generally meant sometime around nine PM.

"It will be quite late when we get back Rick, are you sure you want to go with me?" I asked, not wanting to discourage him, but trying to make sure he was aware of the fact that we would be getting home well past midnight.

"That's OK, Dad," he replied. "I'm looking forward to this night flying."

"Since you have your Private power license Rick, why don't you make up our flight plan for this trip down to Beckley?"

"Great," he replied, as he dragged out my maps for the flight. "Say, Dad. Isn't that the airport located smack in the middle of the big mountains of West Virginia?"

"Sure is, Rick. But that's no problem. They have a good instrument approach there, with a brand new runway and night lighting system just recently installed. Besides, we have a set of two big lights on the King-Air's wings to help us out, and two on our nose gear," I replied.

"Both legs of this trip will be night flights Rick, and I think you will get a real appreciation of how nice this particular ship flies. The weather will be VFR (Visual Flight Rules) so I'll let you be the Captain and fly it as much as you like."

"Gee, thanks Dad, I really appreciate your offer," said Rick.

The company's guests finally arrived and we departed for our flight back to Beckley, West Virginia. The entire flight was as predicted. A very nice calm, smooth and beautiful star-lit night.

Our departing passengers without exception, complimented both of us on the very super-smooth ride they just had.

Dick Barnes made it a standard practice to always wait for

me to get airborne before he left the airport. However, tonight, since it was so nice, he decided he would leave right away.

I cranked up the jet turbo-prop engines, and we were soon taxiing out to the active runway. With no traffic reported, I had planned to take off to the North. There was little or no wind blowing and it was quite late in the evening.

I flipped on all of my lights and this lit up the runway area and it looked almost like daylight. "OK, Rick, are you ready?" I asked.

He nodded his head, and at the same time said, "You bet, Dad," and I moved the power levers to the full forward take-off position. We were both anxious to get airborne again to enjoy the beautiful fall night.

As I started our take-off roll, we picked up speed rapidly. It was about the time I started to rotate, when Rick shouted, "Hey Dad, what's that dog doing on the edge of the runway? See . . . off to our left?"

"Dog? What dog? Why . . . that's a deer, Rick."

"Deer! Deer! What deer!" yelled Rick.

"It really looks like a dog, but believe me Rick, it's a deer," I replied.

It was about this time that the deer started for our lights on the King-Air.

"Look out, Dad!" yelled Rick.

It's too late. There is no possible way I can miss hitting this animal, no matter how I tried. My speed at this point was too slow to rotate and fly, yet, too fast to try and stop.

I could only hope I might be able to pull the nose up and fly up over this charging deer.

No way! The deer hit the left prop and then my nose gear lights. Suddenly, we felt the ship gently drop its nose as the nose gear was snapped back by the deer's impact.

The gear horn started blaring immediately, indicating to me that my gear was not down. All of the lights were flashing red and the warning panel lights glowed brightly, telling me something was drastically wrong.

My landing lights were now shining straight down at the ground as the ship careened crazily down the runway, sideways, towards the edge of the mountains.

I yelled at Rick, "Tighten up your seat belt and harness! I only hope that I can get this ship stopped before we slide off the runway and into the valley below!"

This didn't help Rick's nerves a bit. "That's OK, Rick," I added in a calmer voice. "We'll make it. I think I can get it stopped."

We're still grinding and sliding down the runway sideways. The flashing lights and the loud warning horn made it extremely hard for Rick to hear or understand me.

"Rick," I said. "I can't see any sparks at this time, but I'm sure we must be creating a real fire-storm with this huge 9600 pound metal vehicle, grinding itself to pieces on the concrete runway."

Finally this heavy sled ground to a stop, just inches away from the edge of the mountains, and real disaster. We both took a big breath, and sighed. We can only sit there quite shaken, stunned, and shocked, as the horn and lights are still blowing and blinking.

"Dad!" shouted Rick. "Let's get the hell out of here before it blows up and burns."

"It's all over now Rick, just relax," I said, as I looked over my ship and realized just how lucky we really were that we didn't get killed.

"I have to make a call to my Air Force Operations and advise them of our situation," said Rick.

"I think that call will have to wait a few minutes Rick until I advise the airport of our predicament," I replied.

"I understand, Dad," said Rick.

"Rick, I think an Airliner is due in here before too long and we have the only runway blocked."

"So, what can we do about that, Dad?" asked Rick.

"The only way would be to call them and advise them of our situation."

"Just how can we do that Dad? Everything is out of order now, right?"

"I don't know, Rick. I need to turn on our main power switch and use our radio to let them know."

"But . . . what happens if we have a busted fuel line, and you key the radio and it creates a spark, and . . ."

"Whoa, Rick don't get excited." This was extremely hard for me to tell him this. Outwardly, I was calm, but inside I can feel my heart beating rapidly.

"We have just survived a very bad accident, and I need to advise the operations at the Beckley Airport and let them know of our problem. The only way I could do this right now is to call them on the radio," I explained to Rick.

"I wouldn't be a bit surprised if we didn't puncture a fuel cell, as we slid down the runway on our belly."

Punching on the main switch, I picked up the mike and with baited breath, I called the base operations hoping we would not create any spark with this transmission.

"Beckley radio, we have a problem on your main runway. We are disabled."

My worst suspicions were confirmed when Beckley came on the air and said, "I don't know what your problems are, but you will have to get off the active runway right away. We have an airliner due in here in about forty-five minutes." Then he added, "Stand-by. I'll try and get someone out there to help move your ship off the runway."

I snapped off our radio and master switch and we sat in complete darkness hardly saying a word to each other.

Finally I said to Rick, "Rick, I doubt if they could even budge us."

It was about this time a car came up to our ship with its red lights blazing and siren screaming.

"Now, why in hell do you think he did that?" I asked Rick. "We're the only ship, or people around here for miles."

"Probably to impress us Dad. Or maybe, this is the first time he had to go out on a rescue mission."

The tail is raised up almost 20 feet and the door was about 15 feet above the ground.

"Whatever you do Rick, don't try to jump. It's a long way down to the ground, and it would be a sad thing if we got hurt trying to get out of this ship now."

"Right you are, Dad, I don't intend to go back on active

duty with the Air Force on crutches. I would be hard-pressed to explain that to my Commanding Officer."

We found a rope in the back and were able to lower ourselves to the ground below.

The airport policeman remarked, "Boy, you guys did a lot of damage to your ship."

I repeated his words, "What do you mean? We, did a lot of damage to our ship?" . . . It was a deer that hit us that caused all this."

"Naw, you're wrong, I saw that deer, and he was jumping up and running out to the woods with a bunch of other deer when I drove by."

"No way!" I said. "This craft weighs 9600 pounds, and that deer just couldn't do this, and still be alive.

"How about giving me a ride back down the runway to the place we made contact with the deer, and you and I can check it out?"

"OK, why not," he said, "Hop in."

Both Rick and I hopped in his car. Meanwhile, he radioed to his base and told them that they will have to close the field, adding, "While you're at it, notify the FAA that the field is closed because of this accident."

Using his car's searchlight, we started driving back down the runway. I encouraged him to follow our skid marks to the site of our first impact with the deer. At this point we stopped, and got out of the car. "I'm telling you guys, I saw your deer and a couple of others, jumping down the side of the runway, and off into the woods."

Brushing aside his remarks, I used his car's searchlight, and started scanning both areas alongside the runway.

Suddenly, the bright light picked up a brown outline of something about twenty feet off the runway and on the grass.

As we focused in on this area, our lights picked up what remained of a beautiful, but very dead, deer.

It's hind rump had a huge, deep slash mark it had received from the propellor strike. Closer examination showed its neck had almost been severed.

With this evidence of our accident cause confirmed, the

The King-Air N-5115-D and the damage that resulted from the deer that struck the ship's nose gear on take-off during a night flight at the Beckley, West Virginia airport.

immediate attention shifted back to the King-Air. "We just have to move your ship off the runway before the Airline Flight arrives," said the cop.

"We agree, but how?" said Rick.

"The only way I see to do this would be to get some weight in the tail compartment. We might get the ship up off its nose if we're lucky," I replied.

The policeman said, "I know where we can get a bunch of old batteries for weight. I'll see if we can get them down here to you right away." He departed for the operations with his red lights still flashing.

"Thank goodness he decided that he no longer needed to use his siren," I said to Rick.

He returned in about five minutes and said, "We have about fifty batteries in our junk pile."

God, I hate to think about the possibility of acid leaking out all over my ship. "What if there is acid in them?" I asked.

"No way, they've dried up long ago, and have been stored for years," he replied.

"OK, bring them here and we will give it a try," I said.

With that, he roared off back to the hangar to pick up the batteries.

This time he returned with a truck and a helper. We all pitched in and loaded the batteries as far back in the tail compartment as we possibly could.

Now, even with the weight of five people in the tail, and the fifty batteries, the ship doesn't even budge an inch.

"I have a rope in my truck," said one of the guys. "Let's tie it on the tie-down-eye under the tail, and pull it down with the truck."

"Not on your life!" I said quite forcefully.

"You'll rip off my tail doing that."

"But sir," said this one guy. "We just have to get your ship off this runway within the next ten minutes. We have a scheduled airliner due in here at that time."

"Wait a minute, I've got an idea," I said. "Since my engines and propellors are ruined, why not take those big tie down straps you have on your truck and tie them around the lift jacks on your truck, then tie the props. I believe you can use the props to first lift, then drag my ship off your active runway."

"Good idea," they said, as they followed through with my suggestion. By using this method, they were able to first lift, then drag the ship off to the side and onto the grass area, far enough off the active runway to enable the airliner to use the runway.

It was rather late, close to midnight, but I made a call to Don Bachner, our other company pilot in Columbus. I requested he fly down to Beckley in our other company airplane and pick us up. I warned him all about the deer that were roaming around this airport area. He flew down and arrived about 1 AM in the morning.

I notified our Insurance Company and they thought it would be best to junk the King-Air, due to the location and no maintenance available.

I called my maintenance crew in Toledo and they thought differently. They flew down the following day and inspected my crippled ship.

Three days later they announced they had two rental engines installed, two props, and a ferry permit that would enable me to fly the King-Air back to Toledo for the extensive and expensive repair work. It was almost like a miracle. The King-Air was back in the air two months later.

The deer was estimated to weigh about 140 pounds and the price? $664.28 per pound for a total of: $93,000.

Chapter 24
Two for One

Don Fischer and I were on a routine flight from Columbus, Ohio to Jacksonville, Florida in a Lear 23. We opted for 43,000 feet and as a result picked up a nice tailwind. Our speeds averaged over 480 knots, and we made this trip in the rather good time of one hour and thirty minutes.

It was a very uneventful flight above all the weather and in the absolutely blue, blue sky. On occasion, we would first see a contrail and then pick up a visual on an airliner several thousand feet below our flight path.

Our flight plan called for a quick turn around and return back to Columbus.

After landing and de-planing the passengers, Don ordered the fuel load. Since we are quite light (only 2 of us) we took on 4,200 lbs. of fuel. Fueling completed, I noticed what appeared to be a venting of fuel under the fuselage. I called Don's attention to this saying, "Hey Don! Does this ship always overflow fuel after it is full?"

"Not usually," was Don's remark, as we both got down and looked under the fuselage at the growing puddle of liquid.

"Damn, this isn't fuel," Don said. "It's hydraulic oil."

Don was also a qualified mechanic, but didn't have all his

tools with him. What few tools he had were put to good use in opening up the various panels to see if we could spot the leak. Further checking indicated that we would need help.

Don decided that we would have to call in another mechanic for help, including all the needed extra tools. I went over to operations to call the mechanic. I was shocked to find it would take an additional $100 plus his fee just to get this guy to come out on the week-end. We found out that this was the usual overtime rate charge. I reported back to Don and advised him of the costs.

"OK," Don said. "We have no other choice. Tell him to come on out."

It took the mechanic about an hour to get to the airport. Meanwhile, Don had part of the panels removed, but still needed to take off more. Now equipped with the proper tools, the job was soon completed. All lines, fittings, clamps, etc., were wiped off and checked. Still no sign of the leak. Don decided to run up only the one engine in order to check the hydraulic pump. He exercised flaps, spoilers, and all the hydraulic systems, hoping that it would show up in a leak.

After shutting down the engine, a thorough check was made by both men of the entire hydraulic system. Still no evidence of any hydraulic oil leaks.

The Lear had been in the shop last week for its regular inspection. One passing thought given consideration was maybe some of the hydraulic fluid could have been spilled when the new hoses were installed. Buttoning everything back up, the mechanic added a little bit of hydraulic oil to the reservoir.

Five hours later, Don and I got back in the Lear and I called for the Instrument clearance. Don started only one engine as I got the clearance. Below flight level 41-0 this ship had a voracious appetite for fuel. It was standard practice to start one engine and taxi out with only one engine running. This procedure was very critical as it saved considerable amounts of fuel. Incidentally, this model Lear was well-known as a big fuel gulper, if operated below 20,000 feet.

We both ran through the checklist and by the time we reached the take-off runway he started the other engine and we

The early model Lear 23 pictured here is similar to the Lear 23 described in this chapter.

were ready to go.

With clearance received we took off and climbed out at 6,000 plus feet per minute. Climbing through 24,000 feet we experienced a loud explosion accompanied by a vibration that violently shook the ship. Dust filled the cockpit area as the Lear shook. My heart skipped a few beats and all I could think of at this time was, if I had to die, I pray it would be quick.

We visually checked over the ship and almost immediately found the problem. The hydraulic gauge showed zero! It was quite evident that the hydraulic pump had failed. There was no hydraulic pressure indication on the dial.

We could see where the main landing gear had automatically dropped down, thus causing the vibration. Oh God! The nose gear was still up and in the locked position and no hydraulic pressure in the system to get it down.

I called Jax Center (Jacksonville, Florida) and advised them of our predicament. They cleared us back down to 12,000 feet, and then switched us over to Jax Approach Control.

Jax Approach Control came on the air with, "We understand you have an emergency. Please advise as to the number of souls on board and fuel status."

"Roger, two people and 4100 pounds of fuel."

The radio was silent as we leveled off at 12,000 feet. Then Jax Approach called us back with this comment, "Lear 81 Whiskey Tango, do you need emergency equipment?"

"Affirmative, we only have our main gear down. No nose wheel, flaps, spoilers or brakes, or hydraulics."

"Roger, do you have reverse thrust?"

"Negative," I replied. "And we can't dump fuel."

We had enough daylight left and both Don and I took the emergency checklist out for the third time to re-check our figures. "If we land with 4,100 pounds of fuel we will need 155 knot speed and over 10,000 feet of runway," Don remarked casually.

"Boy, that's a hell of a lot of runway and, according to our charts, the runway at Jacksonville is only 8,100 feet," I replied.

"Don, according to my calculations we would have to land at Cape Canaveral with this load of fuel we are carrying."

"You're absolutely right, Mark. But, by the time we get there it will be dark, and I sure as hell would hate to make an emergency landing in the dark."

"Roger on that Don, I agree with you 100%."

Don then shot back with, "Mark, I think our best bet would be to fly around and at a lower altitude and burn off our fuel so we can land at a lower speed, and a shorter runway length."

I again rechecked our charts and found that if we got our fuel load down to 3,000 pounds of fuel, our landing speed would then be 125 knots and the runway length less than 7,000 feet. Still fast, but not like 155 knots.

The decision was made. I called the Jax Tower to advise them, "We can't dump fuel, so we will plan to circle for about a half hour, in order to use up fuel. This should put us around 3,000 pounds."

The next call from Jacksonville approach control indicated a bit of anxiety in his voice, as he said, "What is the status of your gear now?"

"No change at this time sir, we still have no nose gear, flaps, spoilers or brakes," I replied.

We were cleared to keep circling to our right and told to keep them advised.

Turning back to our emergency checklist, we went through the procedure again, word for word. Meanwhile, my mind was racing ahead wondering if we could get stopped in 8,000 feet. I wonder how bad would we get hurt, if we could not get the nose gear to extend? It was tough, but I had to drive this type of thinking from my mind.

We did have an emergency air blow-down procedure for the gear but, we hadn't used this yet.

We could only pray we would have enough air left for the brakes after using the emergency air for lowering our nose gear.

Don decided now was the time to pull the emergency air bottle lever. He solemnly said, "Here goes."

A resounding thud, and we see the green light appear as it replaced the ominous red light. The indicator had the nose gear down and locked. However, red lights were still on, which told us our gear doors were hanging down in the open position.

Jax Tower repeated their call requesting, "What's your status now? When do you intend to land? By the way, we have all the fire trucks and rescue in place waiting for you."

"It looks like about 10 minutes from now and we will start in," adding, "We now have all three landing gear lights indicating down and locked."

"Hey Don, by the way, when was your drag chute last repacked? and by whom?" I asked.

He replied, "EJA about a month ago. Sure hope it works, as we could have some real trouble if it doesn't."

Finally, our fuel burn was down to what the book said was right for our runway. As we banked towards the airport we could tell the control tower people were very concerned about our current predicament, and so were we.

"One last call and that's it," I told Don. "I don't think we will have much time to talk and tell them about anything as we get closer."

"Roger," he replied, as his eyes narrowed and he put all of his concentration on flying this crippled Lear.

"Jax Tower, Lear 81 WT on final, 2,900 pounds of fuel, no spoilers, no flaps, very little brakes. We have 50 pounds of emergency air left."

"Roger," they replied. "Good luck."

I know this was a crazy question to ask Don at a time like this, but I had to ask just the same. "Say, Don, tell me. When you had the chute packed, was it raining?"

"What the hell kind of a question is that, Mark?"

"I was wondering just how long this chute would take to open, that's all."

My mind was racing at a wide open throttle. I could just envision us rushing down the runway with no way to stop. I could picture us going off the runway and through the fence that surrounded the airport. No question about it, we would have a front row seat to the whole show.

Red lights were flashing all over the airport as we began our approach with our new bug speeds set. (This is a little dial indicator that a pilot sets for his approach speeds and it's actually called the "bug speed".)

I called out to Don, "Plus 4, Plus 6, Plus 10, now minus 5, minus 4, Right on . . . Steady . . . Right on."

Don said, "When you pull the emergency chute handle, be prepared for a big jolt."

"Roger," I replied.

"OK . . . Pull . . . Now!"

There was a loud bang and we both lurched forward in our seats against our shoulder straps.

The chute held, and we came to an abrupt, but safe stop.

When we released the drag chute, the fire trucks picked it up and followed us to the parking area. We both breathed a big sigh of relief.

The fireman in charge, with his silver fire suit on, had a big smile as he came up to us and shook our hands. "You fellows really put on one hell of a show. It was just like in the movies."

"Yeah, sure," we both said smiling, and trying not to show how shaky we are.

Securing the ship, we completed our official report to the local authorities. Don and I, both made for the terminal to see about getting home on an air line.

I got to Peoples Express counter and wound up on their waiting list. Departure time came and I was fortunate to find myself scheduled on board this flight that was to go to Newark, N.J. then on to Columbus, Ohio. It was scheduled to arrive in Columbus around 3 AM the next morning. They were packed to capacity as many of the passengers were college kids on the way home from the spring break.

I had a pretty young girl sit next to me. Ah . . . this should help make the trip more pleasant, I thought.

Taking off, I noticed that we were not gaining attitude, in fact, we appeared to be flying level at about 2-3,000 feet. Not good I thought, but dismissed this as I talked to this girl seated next to me.

"Strange? We should have gone up through those clouds by now," I mumbled to her.

Looking out of the window I hoped to be able to see the stars. The only thing visible was the lights of the town of

Jacksonville with the overcast still above us.

Before long, it started getting hot in the cabin. Everyone was reaching up trying to get more air into the cabin.

Suddenly we started a turn towards the airport. We had been airborne about ten minutes, and I was starting to feel just a bit uneasy.

The girl seated next to me noticed this and asked, "You did tell me you're a pilot, right? Is everything all right?"

I tried to reassure her in a firm voice. "We may have a problem, but don't worry."

After looking out the window, she turned to me again and said, "Tell me, how come our right wing is smoking?"

I looked out and said, "That's not smoke, it's fuel they're dumping."

"What does that mean?" she asked.

Not wanting to alarm her, I tried to put on a good front and be calm saying, "We may have a slight problem."

"What do you mean by that?" she replied, showing just a bit of fear.

"Don't worry, this ship will be OK," I said in as strong a voice as I could muster. About this time, the Captain came on the air and said, "Ladies and Gentlemen, this is your Captain speaking. We have a problem with an engine, and I had to shut it down. We are returning to the airport."

Several women moaned and kids were crying now due to the heat and lack of fresh air. You could feel the tremendous tension throughout the cabin as the Captain had to shut off the lights and air. Everyone waited, expecting the worst.

Banking towards the airport, we could see all the red lights flashing on the ground below.

"This is incredible. A repeat performance."

"What did you say?" said the girl.

"Nothing, nothing really," I replied, but all I could think of was please, keep up your speed. Whatever you do, no steep turns, please!

Finally the runway came in sight and we touched down rather fast and hard.

A thunderous ovation was heard throughout the cabin as

all the passengers breathed a big sigh of relief. The lights came back on, but still no air.

We all had to wait about ten minutes while the emergency equipment and the silver suited rescue team checked us over.

Departing from the ship, we were told that the Captain had detected a fire in one engine and had to use the fire bottle to put it out. The flight attendant said, "We will be back in the air shortly after they clean up the engine."

I assured her, "No way, gal. I fly a jet and if a fire extinguisher is used, the engine must be overhauled, period." I contacted the local motel for a room, ordered two big drinks and planned to call it a night.

Two for one day was a bit too much, even for me.

About the Author
Mark A. Savage

Ever since the age of reason, Mark A. Savage has always wanted to fly. It was almost an obsession with him for as long as he can remember. Perhaps it came about from the early days of his youth, when he fell out of his parents' two-story house. Then again, it could have been the times his uncle as an Army Aviator flew over his home in Rye, New York with the U.S. Army Air Corps P-26, "Pea-shooter."

Whatever the case, it was only natural to continue flying after he had served four years in the USAF. First, as a fighter pilot completing 35 combat missions. Then, as a Ferry pilot, and finally a Transport pilot, during his four years of active duty service.

It was a very fortunate day when he met his future wife Ruth, at Kent State University in Ohio. Ruth had also served in the Armed Forces during WW II, as a Marine. They have made their home in Columbus, Ohio and have been blessed with five children, four boys and one girl. All have grown up now and three out of the five are married. Their oldest son has two boys and one girl, with twin girls that arrived in April of '90.

With the exception of the four years spent at Kent State University, flying has been the major part of his life. He has flown about every military airplane manufactured during WW II. Just about anything that flies has held his interest. Airplanes (dozens of civilian models), balloons, gliders/sailplanes. Turbo prop-jet and pure jets, have been flown and pilot time logged in them.

During the past ten years he has actively flown the Beechcraft King-Air turbo prop-jet over 4100 hours. In addition to this company airplane, he has also flown the other company ship known as the Piper, "Panther."

For a short period of time he flew in a Lear Model 23. Although combat flying was quite exciting, the excitement has not ended with the war. As you read through this book, you will find most of his short stories filled with humor, excitement, and some very dramatic moments that will definitely keep you interested to the very end.

This book was written mainly to entertain and hold your attention. The writing style is quite similar to his recently published book called, *Those Were The Days*.

Whether you like flying or not, there is no doubt you will thoroughly enjoy the many short stories in this book. You will feel that you're right there with him, as he recalls in vivid detail, some of the outstanding flights and real adventures that have made up a major part of his life after the war. He has been flying airplanes since 1940 and has accumulated over 18,000 hours, both in civilian and military aviation. Only 1800 hours were spent flying in the military, and the rest in civilian aviation.

In addition to flying corporate, he has been instructing, building, and flying gliders/sailplanes. He has taught three of his four boys how to fly. They all had their pilot's license before their driver's license.

So, fasten your seat belts and come along with him, as he takes you on many interesting and true, aviation adventures.